STORMWATER
RETENTION
BASINS

Stormwater Retention Basins

Coordinators

Jean-Michel Bergue

and

Yves Ruperd

Ministry of Public Works

France

A.A. BALKEMA / ROTTERDAM / BROOKFIELD / 2000

Aidé par le ministère français chargé de la culture.
Published with the support of the French Ministry of Culture.

Translation of: *Guide technique des bassins de retenue d'eaux pluviales* 1994,
© Technique & documentation, Paris.

A.A. Balkema, P.O. Box 1675, 3000 BR Rotterdam, Netherlands
Fax: +31.10.2400730; e-mail: balkema@balkema.nl
Internet site: http://www.balkema.nl

Distributed in USA and Canada by
A.A. Balkema Publishers, 2252 Ridge Road, Brookfield, VT 05036-9704, USA
Fax: 802.276.3837; e-mail: Info@ashgate.com

ISBN 90 5410 800 2

TABLE OF CONTENTS

Introduction

(For orientation of the reader to intended context)

The guide discusses retention basins of stormwater from urbanised areas as defined by the following criteria:

Water: The water stored is the stormwater

Objective: The objective is basically control of the impact of urban stormwater. This impact may be quantitative in nature in terms of storing the stormwater to suppress or preclude hydraulic shortage in downstream regions. It may also be qualitative in nature when the stormwater carries some pollutants and needs treatment before allowing it to flow into the natural environment.

Network: The retention basin is a structure placed in the stormwater drainage network of a separate system of sewerage even if this separation is not perfectly achieved. It should therefore not be confused with the storm basin situated on a combined sewer network.

Dimension: Two limits have been chosen:
- the upper limit corresponds to basins whose dykes are subject to regulations concerning dams (French bylaw of the 30th December, 1966 and circular of the 21st June, 1967),
- the lower limit is constituted by the other technical alternatives for storage such as the storage on terraces, in ditches, water meadows etc.

The objectives of the guide are:
- to reach a large public from the technicians of urban sewerage to developers and urbanists, and including project supervisors, technical personnel of the community, decision makers, students of hydraulic science and technology etc.
- to propose a rather brief but well-structured document,
- to give as many practical indications as possible regarding the state of knowledge and to present illustrative examples.

There is a large diversity of modes of reading corresponding to the equally diverse nature of readers. It is difficult, rather impossible, that topics selected here will satisfy everyone; their explanations should be of help to the reader, however.

The selection of some topics may appear arbitrary but there is always interaction and on-going research for attaining a better compromise among them.

The distribution of topics into chapters on design/planning, maintenance and management is admittedly sometimes delicate. For example, problems of pollution are involved both in the design (characterisation and dimensioning of the basin) and operation (mechanism of treatment, dimensioning and application of solutions outside the basin).

The distinction between open and underground basins has been explicitly included in the structure of the guide, though sometimes it is explicit in order to avoid repetition.

HISTORICAL SURVEY

Why retention basins? Although to-day this question appears to be incongruous, it is necessary to recall that hardly twenty years ago it was not. This technique, as well as others, has only recently been included in the range of solutions available to the designers of sewerage systems; it has certainly found a more prominent place than the earlier techniques.

The modern sewerage system, initiated in the nineteenth century in western countries, is based on an important principle derived from the hygienistic doctrine: drain away water as fast and as far away from the inhabited zones as possible. Throughout the nineteenth century and now in the twentieth this principle has been the basis for building underground sewerage networks, at first combined and later separate; this continued even into the 1960s.

The end of the Second World War saw the beginning of population exodus to cities and the percentage of urban population increased significantly—from 50 to 75% of the total population. One of the ways to tackle problems of urbanisation consists of creating new cities. Such growth imposes greater concern for developers and urbanists to ensure proper water drainage from these modern cities.

Very logically, the old towns were mostly situated close to streams or rivers; in contrast, the modern cities often occupy plateaus distant from natural outlets.

The mass-plan does not pose a major problem, because as per the current custom, sanitary or drainage engineers are tardily consulted only when the financial estimates of development operations are being prepared and no wonder that often the costs involved present a great surprise.

For classic sewerage systems, the cost of construction of the necessary channels whose length may extend to several kilometres, for draining away the stormwater, may significantly exceed the sum originally assigned for the infrastructure and thereby endanger the feasibility of the proposed new cities. It is for the first time in modern history that the drainage of stormwater is being realised as a major constraint for urbanisation.

The solution to this problem has been readily found: necessity is the mother of invention! Basins are constructed to collect stormwater while it is raining. They are subsequently emptied at a low discharge to natural outlets. Of course, the usual know-how of drainage engineers is disregarded in this process but studies have proposed adoption in France of the Dutch ideas of urban storage tanks, long used in that country because of her topography. Besides, the detention ponds or wet ponds are likely to harvest stormwater and make it available for use in urban areas which may constitute a strong argument for attracting people in new cities at a time when environmental considerations are emerging. Lastly, retention basins enable easy adaptation for stage-wise development of the urban area as they can be planned as soon as the basic real estate has been identified.

The French interministerial circular INT 77/284 (1977) concerning the draining of built-up areas recognises this development by devoting a chapter to retention basins. A breach has opened in the principle on which classical draining techniques were based. Since their introduction in 1977 in France, the design of such basins has evolved significantly from just storage basins to the notion of harvesting stormwater followed by techniques of infiltration. Broadly speaking, this amounts to having at one's disposal a panoply of solutions regrouped under the generic term of alternative techniques for stormwater drainage networks. Apart from the fact that these techniques are all based on the principle of storage, they have the important characteristic of being entirely integrated in urban design. In other words, their execution necessitates a collaboration between the designers of drainage systems and city planners well before undertaking the project. Furthermore, they can be practically implemented in a zone of a separate drainage system only.

Let us cite another development that has taken place during the last few years and is important in this context, namely utilisation of hydrological and hydraulic models in the form of computer software which has become more and more user-friendly and comprehensive. Models are helpful in verifying the feasibility of integration of retention basins in the general system of drainage and greatly facilitate their computation.

What are the important points of this historical survey? At least two ideas clearly state the intentions of the authors of this guide.

First, the **technique** of retention basins is of **recent** origin: it was practically non-existent about thirty years ago. Of course, advancement of knowledge will lead to evolution of this technique. Perfection on adaptation in accordance with the circumstances will bring about clarity of the concept. Let us not be carried away by believing that stormwater basins constitute a panacea for all problems of drainage.

Second, the retention basin is only on**e among other alternative techniques**. It can be replaced—often advantageously depending on the case—by reservoirs, terraces or even ditches along road shoulders. In all cases the choice is to be made by the developer well before execution of the project.

To satisfy the objectives of a specific project every stormwater retention basin is unique. Nevertheless, two standard possibilities of classification can be clearly identified.

Classification According to Appearance:
— open basins
— underground basins.

Classification According to Functions and Usage:
— control of stormwater flow
— remediation
— ecological reserve
— provision of facilities for leisure and recreation.

1   TYPES OF BASINS

1.1   Open Basins

Open basins can per se be subdivided into three subclasses: wet ponds, wet land basins and dry basins.

Detention ponds contain water permanently. Their depth, at least at some points, is sufficient to preclude infestation by aquatic plants at the bottom. During dry seasons they are generally fed by groundwater.

Wet land basins, employed rarely and generally in zones susceptible to floods, can be considered as a particular form of wet ponds. They constitute a fragile ecosystem and the probability of an accidental pollution in them is very low.

Dry basins, as their name indicates, contain no water except during the storm periods. Their entire volume is thus useable for storage.

In terms of the difference of usage, a distinction is often made between lined dry basins and unlined dry basins. This distinction is not absolute and one can certainly find mixed or compartmentalised basins.

1.2   Underground Basins

Underground basins do not occupy surface land and thus do not compete with other structures for a share of the real estate. They generally necessitate large civil works. As they are constructed at significant depth, in most cases, pumps are required for evacuating the stored water.

2   FUNCTIONS AND USAGES

The primary function of retention basins is protection against flooding.

However, as arrangements for treatment or remediation are also provided, their usage often extends beyond this primary function.

Ever since the first use of stormwater retention basins, concerned technicians have been under pressure to extend their use to purposes beyond those for which they were designed.

Decision-makers, after being presented rosy pictures, cannot help dreaming of ideal ecological development of their city. Misunderstandings often lead to resistance against reserving a part of the city land—very precious in most cases—solely to tackle the problems of stormwater drainage.

Fishermen soon begin to consider stormwater retention basins a part of their domain. Major misunderstandings arise between them and the sanitary technicians, in particular when the stormwater created during a storm is deficient in dissolved oxygen and is hence harmful to fish-life. Some fishing societies complain that stormwater pollutes the basin.

Problems of public safety have also been posed. To resolve them, it was proposed that public entry to the basins be restricted by fencing. Nevertheless, there have been some accidents. It may be emphasised, however, that even for accidents occurring in unfenced basins, the legal responsibility of the management would be limited to cases in which flagrant negligence could be established in terms of lack of appropriate warning to the public with respect to possible dangers, as is required for all public facilities.

Finally, as retention basins became more common, drainage technicians faced new problems of exploitation. While cleanliness of the sewers is a customary duty, maintenance of waterbodies is a new task, involving control of vegetation, removal of garbage, control of water quality etc.

These examples show that it should have been apprehended right from the introduction of retention basins that conflicts regarding their usage would arise. But not many persons took notice of it. It was only after some time that progress was made from a position of open conflict to a meaningful collaboration.

The importance of such collaboration lies in the fact that it prompted discussion of compatibility of different possible usages. To illustrate this point, we shall analyse some usages from this point of view. They were chosen because of their frequent occurrence but do not represent the ensemble of all possible eventualities.

2.1 Control of Stormwater Flows

This could involve two objectives: protection against floods and remediation of waters; the two can be mutually contradictory sometimes.

The major constraint associated with the first objective is the obligation of always having available the volume necessary for storage of stormwater from rain storms while the second objective necessitates a prolonged storage time.

As a corollary it follows that the inflows and outflows have to be so matched as to ensure adequate storage capacity.

Remediation is mainly attained by settlement of the solid material carried by stormwater. Contrary to the requirements of the hydraulic function for drainage, the geometry and material of the basin play an important role in decantation and recovery of sludge collected at the bottom. Details of the structure thus depend largely on the objective fixed for remediation: reduction of the annual discharge, balancing and control of the major annual events to limit the effect of shock on the natural environment and treatment of the less intense storm events to limit the effect of stress.

2.2   Ecological Reserve

Usage of 'ecological reserve' often differs significantly from 'hydrological' usage. The long-term objective of creating a reserve is relatively incompatible with the sporadic short-term storage of stormwater.

Using a retention basin as an ecological reserve necessitates considerable protection against pollution which carries the risk of rapid destruction of the ecological balance and degradation of the environment whose restoration would be lengthy and difficult. The existence of chronic pollution from leakages of domestic and other waste waters during the dry season, is thus absolutely unacceptable. For protection against such accidental pollution, always possible, it would be preferable to provide a treatment facility well before the inlet to the retention basin.

Finally, in certain cases, a bypass for such waste water has to be provided. Are we not far from the objective assigned on priority to a sewerage work? In all circumstances, the immediate environs of the basin should always be the least urbanised and access to it strictly controlled.

2.3   Leisure-time Activities

Retention basins could also contribute to the creation of a pleasant urban landscape and provide amenities for leisurely activities such as fishing, water sport, model making, walking promenades, etc. Bathing, however, should not be permitted in any case.

For wet ponds it is particularly important to avoid intrusion of floating matter. Broadly speaking, maintenance of water quality is primordial.

For dry basins, the problem posed by settleable materials may be the most ticklish. It might be tackled by a settling unit upstream of the basin or by cleansing the deposited sludge after every event leading to storage.

A balance must be found between different usages for recreation and a particular treatment applied to the accesses to the basin for these usages. In the same vein, sufficient attention should be paid to the modes of interaction between the basin and its environment.

2.4   Other Usages

Other uses of retention basins can be contemplated. Among those which have been put into practice, we may mention recharge of the groundwater table or the reservoir used for fire-fighting. Such usages depend on the situation at hand; the same holds for analysing their compatibility.

Photo 1   Stormwater basin of Couleuvrain, New City of Melun-Sénart (France)

1

DESIGN AND PLANNING

Multiplicity of participants and their necessary collaboration justify a preliminary description of the various stages in the execution of a retention basin project. One must examine the preliminary studies, choice of type of basin, its dimensions and the potentialities and constraints of development.

I-1 PROJECT EXECUTION

Duration of a project for a drainage system with stormwater retention basins begins with the basic ideas or plans for proposed urbanisation or observations on dysfunctioning or insufficiency of existing networks and extends to completion of the proposed works. It concerns an approach that may spread over several years and involve many participants. The problem varies depending on whether it is a matter of resolving existing hydraulic disorders or offsetting the effects of new urbanisation.

The success of such a project depends largely on the quality of collaboration among the various participants.

Collaboration on a stormwater drainage system and in particular the choice of retention basins as a solution consists of continual exchange of information and suggestions between designers and authorities on the one hand, whose interest is primarily safety against flooding and secondarily, avoidance and control of pollution, and among the residents on the other hand, whose motivation may lie in additional uses that may be permitted by such development. Although hydraulic protection against flooding is the primary objective that should not be forgotten by the residents, the planners cannot ignore the objections and suggestions that may come from the future user community. The future users may find it difficult to understand why so much money should be spent on facilities which statistically function at full capacity only for a few days once every ten years while they are always available for pleasure, leisure activities and improvement in the quality of life.

To facilitate and keep track of this exchange of information, adequate support staff should be employed. The basin dossier, which will be discussed later, is a highly useful tool in this collaboration.

I-1.1   Stages of Project Execution

Execution of a project comprises several stages and there is always the temptation to combine some to 'save time'. In the case of large projects, there is generally a steering committee.

FORMULATION OF THE PROBLEM TO BE RESOLVED
Such formulation is the basis for further studies to identify appropriate solutions. Development of zones of urban activities entails impermeabilisation of several hectares of natural terrain which would subsequently produce additional stormwater flows. Water-courses or drainage networks can accept only a limited quantity of discharge, which is much smaller than the quantity generated. What is the solution? This stage of reflection is conducted mainly by the sanitary engineers.

DEFINITION OF PROJECT OBJECTIVES
This is arrived at in the course of elaboration or recasting of the principal drainage system scheme. It takes into consideration the competence of sanitary engineers, the objectives of the developers and urbanists, and the recommendations of authorities in charge of management of land and quality of the environment.

EXAMINATION AND SELECTION OF TECHNICAL SOLUTIONS
This stage corresponds to a feasibility study and may extend to the pre-project summary. It comprises preliminary studies and requires participation of all the parties involved because it is at this stage that the other usages of the proposed development have to be integrated. An integrated concept of solutions may be facilitated by a methodology that encourages appropriate solutions [8, 69]. Depending on the situation, this phase is concluded by a study of the impact of different options and consultation with the associated public.

EVALUATION, CHOICE OF FINAL SOLUTION AND PLANNING
This is the stage of consolidation of the entire information collected during the course of the preceding phases. It consists of the approval of the detailed pre-project and authorisation of the relevant detailed planning.

EXECUTION AND FOLLOW-UP
This phase comprises all operations of the implementation of works right up to the stage of completion and handover of the project to the city administration.

FOLLOW-UP EVALUATION OF DEVELOPMENT
Theoretically, this operation has no time limit. It comprises the technical, economic, financial, sociological and ecological evaluation of the

development. It necessitates definition, evaluation and analysis of pertinent indicators of good functioning of development plan and quality of service rendered. This analysis of 'developmental life' should form part of the dossier of the basin. In addition to being useful for management of the development, this analysis is a mine of information for good execution of similar future projects even in zones which may be geographically, physically and humanly far removed.

I-1.2 Parties to the Project

INSTITUTIONAL AUTHORITIES
In this term are included the engineering departments, the land administrators as well as the elected representatives who have the difficult role of reconciling various exigencies.

SANITARY AND ENVIRONMENTAL ENGINEERS
Their role is to study the problem in terms of protection against flooding and all aspects of the broader protection of the natural environment. They should propose a range of possible solutions.

URBAN DEVELOPERS (URBANISTS)
The choice of drainage system, taking into account local constraints, may have a bearing on the feasibility of the overall project of urban development (residential, commercial, industrial or other type). Collaboration with urbanists facilitates optimum integration of the drainage work projected in the overall urban development.

MANAGERS OF URBAN UTILITIES
Irrespective of whether they are the technical personnel of the communes or private development agencies or others, there must be consensus of opinion right from the beginning to preclude difficulties that may arise later during operation and maintenance of the structure and facilities.

BENEFICIARIES
Beneficiaries and residents are very important parties; with proper information they can help to guarantee smooth functioning of the works and other facilities (leisure, sports, etc.).

This category of participants may represent diverse groups whose motivations and interests are sometimes mutually divergent. Among them, the engineers are more concerned with quantitative environmental considerations. Architects and landscapers integrate considerations which are often difficult to quantify, such as aesthetics, emotion, ambience etc. Investors are primarily concerned with returns and administrators with conformity and cost effectiveness.

Lack of collaboration among various parties may result in, at best, higher cost of development and, at worst, an overall failure of the operations. The project supervisor has to instigate and animate collaboration.

I-1.3　Basin Dossier for Execution of the Project

Several years may elapse between the initial reconnaissance of a drainage problem, possible appropriate solutions for it and finalisation of the decision to construct the works. It appears essential to conserve the memoirs of the various stages of the project; these memoirs may be in the form of basin dossiers compiled and maintained by the project supervisor.

In the dossier may be kept all information concerning the general framework of the project (starting from origin of the problem) as well as information about the state of its progress. Such information may include transcription of the exact history of the urban development and its environment.

These dossiers are the basic documents that enable, for example, editing an information bulletin for the community or publishing articles in the city paper. They are also helpful in establishing a liaison among the designers, the project supervisors and the beneficiaries.

I-2　PRELIMINARY STUDIES

This stage comprises four steps:
— diagnostics of the drainage system,
— hydrological study of the watershed and streams,
— shortlisting potential sites,
— evaluation of potential sites and defining the scope of complementary studies.

Choosing the type of basin to be created is taken up after these four steps have been completed.

I-2.1　Hydrological Study

Before every project of urban extension or modification of the drainage system (ensemble of components for collection, transport and treatment of water) **it is necessary to have an overall vision of it including receiver waters**.

I-2.1.1　*Diagnostics of the drainage system*

The diagnostics of a drainage system seek complete knowledge of the state of the system (functioning, constraints and potentialities).

In the framework of a project for a retention basin, the diagnostics should focus on:

— identification of the existing drainage-network making a clear distinction between waste water, stormwater and combined waters,

— geometric and hydraulic characteristics of the network,

— delimitation of water supply basins and their characteristics (situation, topography, etc.) in order to appreciate the problems of afferent pollutant loads and risks of accidental pollution,

— state of collection and transport network, screening of mis-connection and inventory of all dysfunctions observed,

— computation of capacity of drainage networks and of any hydraulic deficiencies,

— intensity of peak flow of different frequencies,

— size and capacity of standard sections and hydraulic works, especially those of the downstream network, if relevant, which define the admissible discharge,

— compatibility of needs when carrying at full capacity,

— deficiencies and disorders resulting therefrom (scale of flooding, duration, damage).

At the end of these diagnostics, the first estimate of necessary volume of storage in the given situation is made.

Thereafter one proceeds to simulation of the hydraulic impact of development of the urban area.

I-2.1.2 *Study of the receiving waters*

This study enables appreciation of the level of permissible quality of the discharge of storm period as a function of the sensitivity of the receiving waters and takes into account:

— objectives and criteria of quality,

— discharge during low flow periods,

— physical, biological, ichthyological and ecological state of the environment.

I-2.3 Shortlisting Potential Sites

I-2.3.1 *Elements of urbanisation*

HISTORY OF WATER IN THE CITY

Research on the history of water in the city may help in choosing potential sites for installation of retention basins. As a matter of fact, urbanisation has often masked zones of natural flow or accumulation of water.

Archives of various water sources and old maps are precious documents; they help in the study of toponomy. Names are revelatory of the environment that pre-existed; e.g. fountains, fontanelles, springs, wells, active and dead ponds, pools, etc.

LAND-USE AND OBJECTIVES OF URBANISATION

Following this historical study the important documents to be examined in search of possible sites are all documents concerning land-use or urban planning like master plan and all maps of land occupation, especially those which can be used in conjunction with Geographical Information Systems installed in many communes.

An investigation in consultation with competent authorities enables projection of future land-use. On the basis of this investigation some areas are excluded and others considered preferable.

LANDSCAPING OF ENVIRONMENT

Study of landscapes and sites, especially of the location of graded sites, helps to orient research. Some sites have better landscaping potential than others.

SOCIOLOGICAL ENVIRONMENT

To the list of aforementioned documents on urbanism, may be added those relating to socioeconomic strata. This aspect is important and often necessitates studies of a sociological nature, since a public enquiry is not always adequate. Whether water is to be reintroduced in the sociocultural environment through a stormwater basin or whether 'natural' ecology is to be restored by means of a treated dry basin located in a thick forest or in a clearing, the arrangement be accepted and supported by the inhabitants only if it matches their expectations articulated earlier or otherwise. Although the hydraulic function is of primary concern, other possible usages should reflect the type of population: the young are rather inclined towards sports while older people appreciate quietude, pleasantness, fishing etc. These 'additional' functions would orient the search towards certain sites in preference to others.

The aspects mentioned above are as objective as subjective and are more important with reference to open basins. The considerations stated below, basically 'technical', are exclusively objective.

I-2.3.2   *Technical considerations*

Common sense dictates that retention basins should preferably be created on sites at low levels, cuvettes or thalwegs, because the rainstorm drainage networks generally function by gravity.

The most recent documents of the following type should be consulted:

— topographic maps, oro-hydrographic maps, plans registered in the land records;
— aerial photographs in panchromatic, even infrared if available, (for visualisation of wet, hydromorphic zones);
— geological, hydrological and pedogogical maps, maps of agricultural land if available,
— plans showing the ensemble of earlier existing surficial and underground drainage networks.

As the objective is to make the discharge generated by urban flooding match with a flow compatible with the capacity of the drainage channels, it is beneficial to locate the retention basins in the environs of surfaces generating flood flows. Maps of land-use enable a quick choice of some possible sites and their analysis on topographic and geological maps enables shortlisting them.

Scrutiny of the above documents enables approximate definition of:
— the contours and topography of the catchment;
— nature of the afferent catchment, land-use zoning, and rate of impermeabilisation (examination of the coefficient of generation of flow which enables a rough estimate of the total area subject to flooding);
— eventual existence of differences between topographic and hydrogeological limits of the catchment area (resurgence or loss in karstic regions). The possible existence of external drains which may flow from outside the catchment into it, should also be taken into account.

Storage of stormwater should be taken into consideration right from the moment when planning of drainage networks is initiated and should be given the same importance as roads and other components of urban development works. Sites most favourable on the basis of topographic and geological considerations should be reserved for retention basins. Due attention should be paid to their installation in the environment at the least possible cost and with full anticipation of problems likely to arise in connection with their management and operation.

Some simple principles may help in orienting attention to documents of the most favourable sites. **The object of installation of works in the bed of watercourses is to control a flow rate which most often is not necessary;** the cost of study and installation is high. Contrarily, works installed beside natural watercourses limit the flow rates and storage capacity to that actually needed for urban development. The works are small in size and the cost minimalised. **Therefore, one ought to choose sites which** are situated laterally with respect to the water courses or **are located in small thalwegs**.

In the new urbanisations it is rare that more than 50%, even 70%, of the soil surface be impermeabilised by building up. Retention basins should be installed in the remaining unoccupied 50% to 30% of the area. This is possible if the unconstructed areas are regrouped at the bottom of the valley.

Once perusal of documents is completed, a visit to the envisaged sites should be undertaken. Such a 'tour' would help in validating and completing the information gleaned from various documents.

I-2.4 Analysis of Potential Sites

After shortlisting potential sites, one proceeds with in-depth investigations to ascertain the feasibility of development.

I-2.4.1 Topographical study

A topographical survey is the first operation carried out on the land to be acquired. It helps in:
— verifying the project feasibility as a function of topographic level of water in the stormwater sewers (before the project and after the proposed development),
— evaluating the storage capacity for the various ratings of water level and installation of a dyke if it is included in the plan.

I-2.4.2 Geologic and geotechnical study

Study of the subsoil should be made to ascertain:
— is the mother rock adequate to ensure good structural stability of a dyke and the works?
— will the works need to be anchored? if so, at what depth?
— will there be risk of erosion of underlying materials in case of accidental infiltration (as occurs in the presence of strata of gypsum or coarse sand);
— how to ensure the stability of an underground basin?
The first step entails perusal of geological documents in order to study structural characteristics of the chosen site and, if necessary, examination of the sites to ascertain whether they are rocky or granular. Depots for storage of extracted materials should also be surveyed.

It may become necessary to carry out supplementary reconnaissance by drawing samples and testing the materials. These operations may consist of:
— reconnaissance of geological strata by boring, drilling, piezometric measurements etc.,
— identification of materials, definition of modalities of compaction, delimiting borrow zones.
— determination of permeability,
— standard geotechnical tests for estimating the cohesion of materials, the angle of friction, etc.,
— non-destructive tests for cavities by geophysical prospecting (microgravimetry), etc.
A detailed plan of reconnaissance of soil, specifically that needed for construction of a dyke, is given in Section II-1.1 [41, 54]

I-2.4.3 Hydrogeological study

CASE OF WET PONDS
To maintain a wet pond one needs:
— either a basin which has been naturally or artificially made watertight,
— or a permeable structure with a permanent phreatic layer present, with the **provision of adjustment of the normal level of the retention**

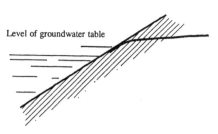

Insertion: groundwater level

basin to the low level of groundwater or under it in order to prevent risk of pollution of groundwater by water from the basin. (see figure)

The preliminary hydrogeological study for installation of a wet pond should:

— determine precisely the characteristics of the impermeable floor of the aquifer (level, thickness, slope, permeability),

— evaluate the capacity of the aquifer and its hydrological characteristics (flow direction and discharge);

— determine the minimum and maximum annual levels of the groundwater table, the corresponding frequency of occurrence and, if possible, evaluate the lowest level for the dry year of a given frequency;

— indicate whether the groundwater has been exploited, especially downstream of the planned basin; if yes, by whom and for what usage; provide knowledge of the water quality and the degree of vulnerability of the groundwater;

— lead to evaluation of risks of modification of the level and the direction of flow of the groundwater if important works such as trenched roads, expressways and/or railway lines are planned.

Absence of groundwater creates problems of feeding and renewal of water in wet ponds. In such cases the bottom and the walls of the basin are made impermeable upto the minimum water level desired in the retention basin, or a dry basin is planned or an alternative sewerage device is chosen (wet land or other alternative techniques).

The choice is not inconsequential since the techniques of implementing impermeabilisation and the related works are expensive. Hydrogeological reconnaissance may suggest intermediate possibilities: implementation of partial impermeabilisation, artificial local raising of the groundwater level (watertightness mask) or slightly lowering the bed of the basin (Fig. I.1).

CASE OF DRY BASINS

A hydrogeological study similar to the one presented above is needed for this type of basin also, in order to estimate risks of accidental pollution of the groundwater and to plan adequate measures for ensuring protection against it.

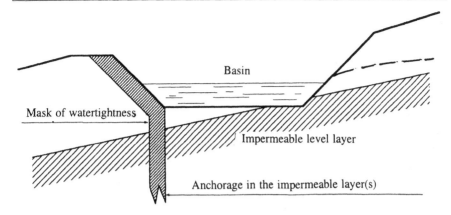

Fig. I.1   Maintenance of water in a basin by watertightness mask.

CASE OF UNDERGROUND BASINS

A hydrogeological study enables precise assessment of:
— effect of eventual presence of a phreatic layer submerging the basin on the design of the underground basin;
— risks related to settling of the adjoining soil in case of lowering of the groundwater level during construction;
— risks due to pore pressures produced by the groundwater;
— quantities of permanent or fair-weather discharge in drains;

SOME EXAMPLES OF FAILURE DUE TO FAULTY INFERENCES FROM THE
HYDROGEOLOGICAL STUDY

A wet pond can be transformed into a dry basin if the bottom of the basin is placed at a level higher than that of the groundwater table.

Determination of the lowest level of the groundwater table requires piezometric information over a number of years. Often, this information is incomplete and it becomes difficult to correctly define the desired level of the bottom of a wet pond.

In the case of incorrect placement of the bottom of the basin above the lowest level of the groundwater table, two major problems may be encountered:
— plant infestation due to insufficient depth of water in the basin; the basin may then resemble a rather vague terrain of fallow land than a pond;
— risk of contamination of groundwater by percolation of polluted stormwater which pours into the basin.

This type of failure, inscribable to the preliminary studies, can also be caused by a modification of the hydrogeological environment of the basin. Problems may also be created by a trench road formed by erection of moulded walls up to the level of the groundwater table. If the stormwater basin is situated downstream of the trench (relative to the direction of flow of groundwater), it could dry up due to interruption in feeding. Conversely, a dry basin with permeable bottom and walls situated upstream of the trench forming an underground barrage could be transformed into a wet land, even a water body.

Photo 2. Wet pond transforming into a dry basin due to lowering of the groundwater level.

It is easy to see from the above why there is interest during the preliminary studies in acquiring knowledge of all developmental works planned for the catchment and also the downstream parts of the watershed.

I-2.4.4 Climatological study

This comprises:
— study of local or, failing that, regional precipitations over a period of observation which is sufficiently long with regard to the return period for which the work is designed (of the order of thirty years for a ten-year return period);
— study of evaporation and evapotranspiration.

The first point relates to all types of retention basins and is dealt with in Section I-4.2. The second relates to wet ponds and is discussed below.

The maintenance of a permanent water body results from an overall positive balance between inflow and outflow.

Evaporation from free water surfaces is determined from the measurements carried out on an evaporation pan (Colorado type or class A) or a PICHE evapotranspirometer. This data, averaged over a month or a day, may be available from a meteorological office such as Météo-France. Diverse formulas are available for calculating it: MEYER'S, COUTAGNE's etc. They are adequate for obtaining an order of magnitude of the loss by evaporation as a function of climate and the mean water depth. In the case of a retention basin having a large area covered by flora, the concept of evaporation is substituted by that of evapotranspiration. There are two measures of evapotranspiration: Potential Evapotranspiration (PET) related to climatic factors (temperature and relative humidity of air, wind speed, albedo or reflective index of the surface, etc.) and Real Evapotranspiration (RET), which is more agronomical and depends on the type of vegetation and its stage of growth. In practice, regional data of the former published by meteorological offices is generally referred to.

For example, for a dry year over a ten-year period of return in the region of Ile de France, the monthly values of PET are shown in the Table below (THOMACHOT, [75]).

Month	J	F	M	A	M	J	J	A	S	O	N	D
PET (mm)	10	15	50	86	123	149	145	130	93	50	22	10

I-2.4.5 Pedological study

A pedological study determines the nature and thickness of soil horizons actually present at the site and their agronomical aptitudes. **One must study the usage being made of the earth to evaluate the risk due to excessive presence of fertilizers and pesticides.** This study enables calculation of the quantities available for reconstitution of soil which could carry a vegetative cover around the basin and **formulation of recommendations regarding the modalities of stripping.** It also determines the vegetative species which could be best adapted for development of the basin and its approaches.

I-2.4.6 Occupancy of subsoil

Preliminary knowledge of the position of all underground networks (drainage, water supply, gas, electricity, telephone, etc.) likely to be present on the site is indispensable. This study is a must in the case of underground basins.

It is not rare that digging of a basin results in discovery of rich specimens of archaeology. There are regulations governing items so discovered and their treatment.

Finally, legal aspects of occupancy and/or sharing of land-surface and buried materials must be studied with care, especially in the urban areas.

I-3 CHOICE OF TYPE OF BASIN

The above preliminary studies will have enabled fixing the location of proposed stormwater basins. The next question pertains to the choice of type of

basin which will ensure the best service. There are numerous objective and subjective criteria for making this choice, some of which are of paramount importance indeed. These criteria can be divided into four classes in remove order of importance:
— physical criteria,
— urbanistic criteria,
— economic criteria,
— environmental criteria,

I-3.1 Physical Criteria

These comprise topographical, hydrological and geotechnical criteria whose preliminary analyses provide factors governing the performance of the basin.

The nature of the subsoil, depth of the groundwater table in relation to the natural terrain, its fluctuations, dimensions of sewers at inlets and outlets (very often prescribed), even technical feasibility of a reinforcement of the network, are the critical factors. They enable making a choice among wet pond (for example if the pit is impermeable and there is no risk of pollution of groundwater), dry basin, wet land basin and underground basin (for example, if the stormwater is susceptible to strong pollution, often the case in the vicinity of road junctions or along the highways).

I-3.2 Urbanistic Criteria

I-3.2.1 Land-use

Two cases may be encountered:
— if the planned basin site is away from inhabited area, situated on land not frequented by pedestrians (road junctions, railway premises, airports, etc.), a dry basin is a feasible choice. In the vicinity of aerodromes, dry basins are particularly preferred for reasons of safety (preclusion of birds, fog etc.);
— if the basin is to be located in an urbanised sector or in a zone frequented by pedestrians but with no facility for recreation, nuisance created must be insignificant detention ponds or dry basins are equally acceptable solutions.

In cases where buildings occupy all or most of the space, underground basins or other techniques of storage have to be resorted to.

I-3.2.2 Nature of relevant catchment area

If the catchment includes industrial, commercial or parking areas—sectors with strong potential for pollution—it is preferable to choose dry basins with provision of treatment of stormwater at the source (parking, exit etc.), or at the outlet of the sewers into the basin.

I-3.2.3   *Resident population*

Residents of the sectors to be drained may influence the choice of location and other 'usages' of the stormwater basins. The type of basin chosen should take their views into consideration.

I-3.3   Economic Criteria

The aims of these considerations are: to **reduce** the intensity of flood by means of retention basins, to help reduce the cost of investment and transport works, to **phase out** and **adapt** the construction of retention basins as the urbanisation proceeds and to **split** the corresponding costs and **lessen** the financial burden.

Construction and management of basins is bound to result in additional expenditure but experience shows that the cost of a drainage system with wet ponds comes to around 40 to 60% of the cost of a traditional system of flood drainage by sewers capable of carrying the entire flood from a storm of once in ten-year frequency. It should not be forgotten that if efforts are made to ensure protection for storms of more than ten-year frequency by increasing the diameter of sewers of a traditional drainage system, **this will only result in aggravation of the situation downstream due to flooding of the receiver stream. Open detention ponds ensure protection downstream even beyond the ten-year risk without involving appreciable additional cost**.

In addition to the above considerations, due attention must be paid to the problems of operation and maintenance associated with choice of basin type. It is important not to lose sight of these constraints of operation.

Often the management of basins is shared by two or more departments. Horticulture departments may be responsible for the development and maintenance of land surface and vegetation including clean-up of the water body while sanitation personnel may manage the hydraulic works and attend to the maintenance of appropriate water quality.

I-3.4   Environmental Criteria

Environmental criteria are playing a bigger and bigger role in the choice of type of retention basin, essentially for two reasons:
— on the one hand, evolution of the attitude of inhabitants towards environmental protection,
— and, on the other, the fact that regulations regarding water pollution and treatment of wastes are being formulated at the national as well as global level.

In this context, too, one may classify the specific impacts of various types of basins as given below.

I-3.4.1 Impact on hydraulic regime

The objective of stormwater retention basins is precisely the **least possible modification in the hydraulic regime of a stream or river used as the receiver body**, so as to leave it in about the same state in which it existed before urbanisation. Retention basins help in buffering the flood flows from the urbanised area and thus restore them closely to the 'natural' state. The only difference is that the emptying of the basin is accomplished over a long period of time.

The following options are available for the choice of the basin best suited for this purpose:

— a single large basin,
— several small basins arranged in parallel or in series.

However, in making the choice of type of basins and especially the arrangement of a number of basins in a drainage system, one has to take into account the possibilities of aggravation of flood surges if a second or third surge should originate concomitantly from urban areas when the first surge occurred due to flows from agricultural or unexploited zones. In such a case one has to very carefully plan the spatial layout of stormwater basins and the combination of wet ponds and dry basins, and provide for dynamic management of the outflow from various basins (cf. [80]).

I-3.4.2 Impacts on water quality

Stormwater 'washes' or flushes the urbanised surface, parking lots and roads. The quality of water which flows into the stormwater retention basins is hence degraded by pollutants resulting from various urban activities (automobile traffic, etc.), garbage, flottants, sediments, dissolved salts, pesticides, etc.

The manner in which the water quality is affected depends on whether it is a wet pond or a dry basin. **Measures that enable maintaining or attaining the desired or specified water quality become the determinant criteria for the choice of appropriate retention basins.**

CASE OF WET PONDS
The wet pond plays an important role. **It dilutes the pollution** and simultaneously results in:

— **sedimentation** preferentially at the inlet to the basin where the speed of water, which is high in the sewers, reduces to almost zero: heavy particles settle and light particles float; thus a purification of the primary type occurs in the basin;
— **oxygenation** of water in contact with the atmosphere and oxygen released by the photosynthetic processes in the basin;
— **odourless decomposition** of organic material under the action of aerobic bacteria;

— **assimilation** by vegetation, of eutrophic compounds; these serve as a source of nutrition from which the plants can fabricate appropriate organic material utilising solar energy (photosynthesis).

The residual pollution flows slowly into the environment thereby precluding the effect of shock.

Although the stormwater basins can 'absorb' a certain quantity of pollution through the aforesaid natural processes, they should in no case be used as a receptacle for untreated waste water or sewage. **Wet ponds should be planned only in strictly separate drainage network.**

CASE OF DRY BASINS

In this type of basin, protection of the quality of the environment is accomplished in a generally less complete manner except in particular circumstances (compartmentalisation, controlled management of evacuation).

— The effect of shock in the course of receiver water is avoided by slow and controlled release of the polluted stormwater. However, dilution of pollution and oxygenation of water take place largely in the receiver waters only.
— With good design and management, the suspended material and flottants are trapped in the basin; their advance in the environment is thereby prevented.

CASE OF UNDERGROUND BASINS

These basins are also helpful in protecting the receiver waters through retention of readily settleable pollutants and staggering of pollution over time.

I-3.4.3 *Impacts on urban landscaping*

Large architectural projects often include a waterbody. Many modern residential complexes, industrial areas, commercial centres etc. utilise the aesthetic upgrading offered by water bodies to increase the value of the buildings and render them more pleasant to live in. Involvement of landscape architects and urbanists in the design of retention basins is thus recommended.

A basin in the form of a canal can create the appearance of a very attractive river if its banks are lined with good quality material, adorned with flowering or other plants and surrounded by promenades for walks and picnics.

Dry basins may be adorned with gardens and/or sports arenas as well as green belts or water channels.

For the underground basins, one should consider integration with the architectural techniques installed on the surface in the area.

I-3.4.4 *Impacts on quality of life*

Integration of sociological factors and design aspects is a recent phenomenon. It is taken into account at the stage of impact study through a procedure during which the public is invited to give its opinion, especially if such evaluation is accompanied by a public hearing regarding the impact assessment.

The presence of water in an urban environment is always a great attraction for the inhabitants. Provision of aquatic gardens in the interior of cities is a major assets for attracting a certain quality of life. Walkways on the banks of watercourses are becoming popular in many cities today.

Wet ponds cannot remain immune to this passion of the community, which may pose some problems but still demonstrates the viability of the idea. **All these criteria of choice should be arranged in the form of a table of multicriteria analysis, with notes and reservations, which can serve as a tool in arriving at an optimal decision.**

I-4 DIMENSIONING OF BASINS

I-4.1 Introduction

Once the site of works has been defined, the dimensioning of retention basins is generally carried out by successive approximations.

The first step consists of estimating the volume and the release rate in accordance with objectives of protection and hydraulic constraints. At this stage one should proceed to the evaluation of costs because they are an important factor in deciding the orientation of the project.

If necessary, as the second step one should consider other objectives and/or other constraints, such as protection of the receiver waters, various usages of the site, such as leisure and recreation, for evaluating the final dimensions of the work or investment costs of its establishment and also the operation and maintenance entailed in these secondary functions.

The first section is devoted to hydraulic calculations and simple methods have been discussed. They are used mainly at the level of the preliminary study for determining the order of magnitude of volume of the retention basin and also for the final design of works of small size. More elaborate methods are needed for confirming the feasibility of large projects.

The second section deals with the problem of remediation of the stormwater. Research has been in progress for more than a decade but the results obtained so far are incomplete. Also, the available experimental data are insufficient for formulating norms or precise guidelines regarding the works to be carried out. Nevertheless, recommendations based on the data yielded by the first set of experiments are proposed.

Lastly, the third section treats the particular case of works of infiltration. Simple formulas are given for computing infiltration in the most common cases.

I-4.2 Hydraulic computations

I-4.2.1 Introduction

The purpose of hydrological and hydraulic studies is to determine the useful storage volume of the retention basin as well as the conditions of safety of its operation.

Let us recall that the volume of the retention basin is the sum total of the following volumes:

— volume occupied (eventually) by permanent waterbody;
— useful volume, or volume available for floodwater storage, often corresponding to once in ten years flooding;
— safety volume which lies between the FRL (full reservoir level, the maximum intended water level in the basin) and the water level beyond which spillage begins;
— additional safety volume between the FRL and the top of the dyke (freeboard).

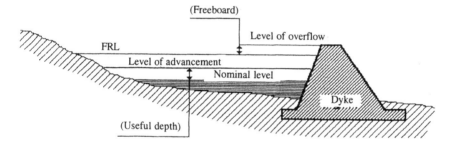

Evaluation of useful volume and safety volume depends on:

— characteristics of stormwater from the project area into the basin;
— discharge characteristics and safe flowrate over the spillway.

Although generally negligible compared to the rate of stormwater from the project area, the rate of inflow from other sources and from phreatic layers should be evaluated at the start for:

— dimensioning the regulation works (rate of inflow from such miscellaneous sources should be taken into account in the release rate;
— estimating the possibilities of maintaining the water level during dry-weather periods.

Further, inflow is also involved in the calculation of the volume to be stored.

I-4.2.2 Evaluation of release rate

A hydraulic study taking into account the ensemble of inflows from the catchment areas situated downstream of the planned work helps in determining the residual capacity available in the drainage channels and, subsequently, the discharges that can be released from the retention basin.

The residual capacity available in the drains is evaluated for a given return period of the inflow and eventually at any instant 't' of flooding.

The inflows from catchment areas situated downstream are evaluated:
— either by methods that give only the peak value of flow: Caquot's* method for the urban catchment area; rational method and Crupedix* method for mixed catchment areas, rural areas and wilderness areas;
— or by means of mathematical models capable of continuous simulation of the storm event. In this case, it is possible to evaluate the residual capacity of the downstream network at every instant of flooding and to determine the opportune moment for regulating the release rate from the retention basin.

Let us recall that the French *Technical Instructor* of 1977 [40]** pertaining to the design of drainage networks for urban areas [40] specifies that 'it is a priori a good management practice to protect against the risk of a once in ten years flood cycle' and that 'in strongly urbanised regions of flat relief, the project supervisor should not hesitate to design the major feeders for absorbing the discharge of 20 years, even 50 years period of return....' These recommendations primarily concern dimensioning of collectors.

I-4.2.3 *Choice of frequency of failure of basin*

The average frequency of overflow due to abnormal flooding is called the frequency of failure of the retention basin. Selection of this parameter is based on an in-depth study of the consequences suffered by the downstream region by additional discharges due to such overflows from the basin.

As a general rule, the adopted frequency of failure of the retention basin is equal to or less than the permissible overtopping of downstream networks. Thus for small catchment areas (up to a few tens of hectares) for which the additional release from the basin does not lead to harmful consequences, the generally accepted frequency of failure is once in 10 years, or 20 years.

For larger catchment areas or when overflow entrains harmful consequences, the frequency of failure is taken as once in 30, 50, 100 years, or even longer.

I-4.2.4 *Weir for Draining floods*

The useful volume of a storage basin is thus primarily determined by the relative degree of concern for urban sanitation, i.e., handling commonplace floods as opposed to rare and exceptional flooding.

In any case, even rare or exceptional floods should be able to pass without causing such damage to the works as might threaten public safety due to their partial or total rupture.

---

* Both are French methods. In English literature they are known respectively as 'rational method' and 'Soil Conservation Service method'.
** This document is rather old and is not likely to be available.

In this context, dykes involve appreciable risk, which with other things being equal, increases with the height of the dyke and the volume of retention. Consequently, these works need to be equipped with a device capable of guaranteeing protection against rupture. For this purpose, the works should be designed on the basis of a hypothetical flooding of low frequency, of the order of 10^{-4}.

During rare or exceptional flooding the major portion of the floodtide is usually drained by an appropriate arrangement of water streets: channels in the direction of the steepest slope ensure drainage of the floodtide while those in the perpendicular direction play the role of holding and distribution of the tide. Consequently, the catchment area which constitutes the basis for calculation of the useful volume of the retention basin may differ from that taken into consideration for drainage of floods of low frequency:

— either because, due to various artifacts, the effective catchment area becomes much larger than the topographic catchment area;

— or because a large portion of the tide is diverted in the direction of other sites.

The simplest case is that of works installed in a manner that only water brought in through the planned network of drainage channels enters the retention basin.

I-4.2.5 How to drain unusual floods?

In principle, the discharge due to flooding that needs to be evacuated is smaller than the peak flood flow of a retention basin because the basin creates an effect of lamination.

For basins bounded by a dyke, a good estimate of safe rates of stormwater is obtained by the largest flood flows that would be carried away by the peak discharge calculated with a water head of around 50 cm over the spillway crest. The capacity of evacuation may increase appreciably in this case with a small additional head. A margin of safety is allowed by the height lying between the full reservoir level (FRL) and the top of the dyke; this margin is termed the 'freeboard'. The rule for small basins is to hold this freeboard to at least one metre.

An alternate solution would be to prevent submersion of the top of the dyke. For this, one must find a watercourse such that the overflow will be bypassed upstream of the retention basin: diversion of water towards another catchment area by means of streets or natural channels for example.

I-4.2.6 Determination of useful volume

The methods employed are based on graphic or numerical resolution of the following set of equations:

(1) Equation of conservation of volume:

$$dV/dt = Q_e - Q_s \qquad (1)$$

where V is the volume of water in the retention basin; t the time; Q_e the rate of inflow; Q_s the rate of outflow.

(2) Equation defining the volume in the retention basin in terms of water depth

$$V = f(h) \tag{2}$$

where h is the height (or depth) of water in the basin.

(3) Equation expressing the rate of outflow from the basin in terms of depth of water in the basin

$$Q_s = g(h) \tag{3}$$

SIMPLE METHODS

These include the 'rain method' and the 'volume method' and are given in the French *Technical Instructor* of 1977 [40]. In these methods it is assumed that the rate of outflow from the retention basin is constant. The solution to the problem relies uniquely on the resolution of the equation of continuity.

Determination of the hydrograph requires the **runoff coefficient 'Ca'**, **which is a measure of the efficiency of turning global precipitation into stormwater, and evaluation of the effective surface area 'Sa'**, defined as the product of the physical surface area of the catchment, S, and the runoff coefficient

$$Sa = S \times Ca$$

The runoff coefficient, Ca, is essentially the ratio of the volume of the stormwater flow to the volume of total rainfall.

The overall runoff coefficient for a large, heterogeneous area can be calculated from the partial runoff coefficients Ca_i of homogeneous zones of surface area S_i:

$$Ca = \left(\sum_{i=1}^{n} Ca_i \ S_i \right) \Big/ S$$

The runoff coefficients used in sanitary projects for a ten-year cycle of rain in an urban environment can be considered, to the first approximation, as given in Table I.1

RAIN METHOD

This method requires knowledge of the 'intensity (i)-duration (t)' curve corresponding to the period of failure of work (T) adopted, say i (t,T).

The curve of specific height of water H (t, T), defined as height of water per unit contributive catchment area, can be deduced from the intensity-duration period curve i (t, T) by the following relation:

$$H (t, T) = i (t, T) \times t$$

Table I.1 Runoff coefficients for ten-year cycle for homogeneous zones used in sanitary projects in an urban environment (after SAUVETERRE [63]) for calculating effective surface areas

Type of land-use	Runoff coefficient for ten-year cycle
Green zones, sports grounds etc.	0.25 to 0.35
Individual housing	
12 family units/ha	0.40
16 family units/ha	0.43
20 family units/ha	0.45
25 family units/ha	0.48
35 family units/ha	0.52
Apartment housing	
50 family units/ha	0.57
60 family units/ha	0.60
80 family units/ha	0.70
Public buildings	0.65
Commercial areas	0.70
Supermarkets	0.80 to 0.90
Parking lots, roads	0.95
Waterbodies	1.0

$H(t, T)$ is expressed in mm if $i(t, T)$ is in mm/h and t in hours.

Also, if Q_s is the outflow rate of the retention basin of effective surface area Sa, the specific outflow rate q_s is given by

$$q_s = \alpha \times Q_s/Sa$$

If Q_s is measured in l/s and Sa in ha, q_s is given in l/s/ha or can be expressed in mm/h , with the dimensionless conversion factor α being equal to 0.36.

The equation (1) of conservation of volume is solved graphically by noting that the maximum height of water D_h to be stored in the basin is equal to the maximum difference between the curves for $H(t, T)$ and V_r against time. To obtain D_h, a tangent is drawn to the curve for $H(t, T)$ at such a point that the tangent is parallel to the straight line representing the $V_r(t)$ curve (Fig. I.2).

The volume to be stored in the basin can be directly obtained from the following relation:

$$V_s(Q_s, T) = 10 \times D_h \times Sa$$

where V_s is expressed in m³, D_h in mm and Sa in ha.

Note: Fig. I.3, derived from Fig. I.2, shows that it is not desirable to provide for water to be stored in two basins in tandem which are emptied at two differently regulated discharges. The total volume to be stored in the two-basin case, B + C , shall be larger than A, the volume to be stored in a single

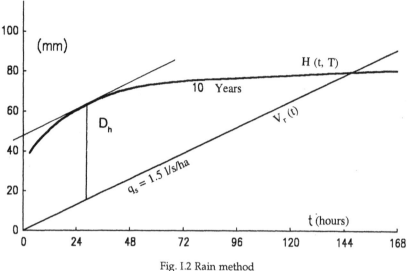

Fig. I.2 Rain method

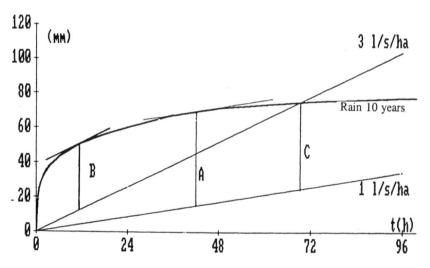

Fig. I.3 Retention in two basins in tandem emptied at different release rates.

basin for a given final release rate. In the case of compartmentalisation of the basins in tandem, discharge of all compartments shall be identical.

The relation between the specific height H (t, T) and intensity-duration-period of rainfall forms a group of envelope curves. A sketch of the type shown in Fig. I.2 can be drawn for each envelope curve and from these, storage volumes for frequencies of once in 10 years etc. worked out. The

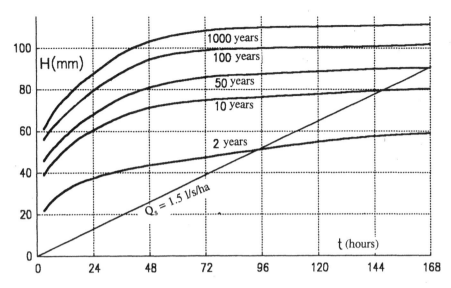

Fig. I.4 Group of curves for rainfall of different frequencies at Paris-Montsouris.

results given in Fig. I.4 show that the stored volume increases with level of protection desired.

VOLUME METHOD

The volume method is recommended by the *Technical Instructor*; it defines for France, three regions in terms of the amount of rainfall.

While in the rain method one starts with a statistical study of precipitation, in the volume method one starts with the same rain data but after deducting the volume corresponding to each value of the various release rates. This involves evaluating the D_h (q_s) for the ensemble of observations and then proceeding to statistical analysis of annual maxima. Actual volumes to be stored, corresponding to several values of release rate, are classified frequency-wise and this classification done for an entire series of rain events. In terms of statistical analysis, the volume method is comparatively more orthodox.

Values of D_h corresponding to four different frequencies (2, 4, 10 and 20 years) have been charted for these three regions (see Fig. I.5).

In practice, one proceeds as follows:

Step 1: Calculation of the contributive area:
$$Sa = S \times Ca \text{ (Sa in ha)}$$

Step 2: Identification of release rate:
$$Q_s \text{ (m}^3/\text{s)}$$

Step 3: Calculation of discharge per hectare of contributive area:
$$q_s \text{ (mm/h/ha)} = 360 \, Q_s/Sa$$

EVALUATION OF SPECIFIC STORAGE CAPACITY
OF STORAGE BASINS

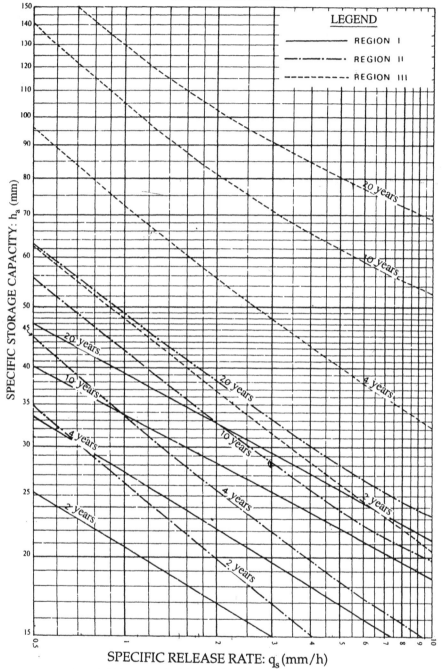

Fig. I.5 Volume method (chart Ab7 of French *Technical Instructor*).

Step 4:   Choice of the period of return, T

Step 5:   Determination of specific height D_h of water to be stored using the chart for the region relevant to the project (see Fig. I.5)

Step 6:   Calculation of volume of storage: V (m³)

$$V \ (m^3) = 10 \times Sa \times D_h$$

The volumes obtained by the volume method are generally larger than those calculated by the rain method by 5 to 50%, depending on the region, release rate and the period of return.

LIMITATIONS OF THESE METHODS

Very easy to use, these methods are employed in numerous cases; they have been formulated on the basis of the following simple hypotheses:

— Absence of losses of precipitation in the catchment area. In fact, there are some losses which become significant if the catchment area is an extended one and the duration of rain is short. Let us remember, however, that rain generally produces large volumes of flow only when it is of long duration.

— Runoff coefficient is constant irrespective of duration of precipitation; in reality this coefficient increases with intensity and duration of rain. Consequently, the results obtained by calculation differ from real values, the difference increasing if the catchment area is only slightly urbanised and the release rate from the retention basin small.

— Discharge at the outlet of the retention basin is constant; this hypothesis assumes that outflows are fully regulated.

The volume method given in the *Technical Instructor* defines only three regions based on the amount of rainfall; the curves given in the *Instructor* may not cover the specificity of rainfall in cities where a project is planned. Also, the curves are limited to a cycle of 20 years.

In conclusion, these methods can be recommended for catchment areas of limited contribution (new development regions) and applied only to obtain the first approximation for catchment areas larger than 200–300 ha in size.

DISCHARGE METHOD

This method consists of simulating the flow by means of mathematical models, each representing one or several stages of the cycle of stormwater. The volume of the retention basin is calculated by numerical resolution of the system of equations (1) to (3) given at the beginning of this subsection (I.4.2.6).

Simulation is accomplished on the basis of:

— either rainfall of given frequency and different durations. This rainfall is generally derived from the intensity-duration-period curves. The period of return thus calculated for the maximum volume is adopted as the design frequency of occurrence of rainfall for the project undertaken.

— or rainfalls actually observed over the catchment area or some other catchment areas whose climatic conditions are fairly identical. Determination of volume of retention basin for a given frequency cycle is carried out on the basis of statistical analysis of the results of computation.

Due to lack of available measurements over a sufficiently long period at several points of the catchment, this approach is rarely employed.

The major criticism of the discharge method pertains to the ability of stormwater models to reproduce the phenomena observed in natural and rural basins. Furthermore, satisfying initial conditions of the state of soil saturation for storm events is not easy.

But its shortcomings notwithstanding, the discharge method is commonly used today. In fact, compared to the 'simple methods', it offers the advantage of a more sensitive study of the global model using different parameters of entry.

Several softwares designed for analysis of sanitary networks aid in determination of useful volume of retention basins: CANOE (SOGREAH-INSA of Lyon), TERESA (STU) among the softwares in France and MOUSE (Danish Hydraulic Institute), HYDROWORKS (Wallingford) among the softwares abroad, may be mentioned.

Figure I.6 illustrates utilisation of this method.

I-4.2.7 Regulation of discharge

Dynamic management of outflow from a basin helps to optimise the functioning of the sanitary system for each storm event with respect to the objectives pursued; these could include:

— for a given useful volume of storage, decrease the frequency and/or volume of overflow to downstream drainage networks;
— improvement of protection of the receiving water by storage and sedimentation of the stormwater in the basin.

Regulation is ensured by means of control devices (valves, overflow over an adjustable weir, pumping stations, siphons...) whose setting is adjusted as the event progresses [36, 51, 80].

I-4.3   Remediation

In urbanised/inhabited areas, provision of treatment of domestic and industrial waste water is made leading to improvements in quality of the stormwater and impacts of pollution hence become visible only in open spaces downstream. Pollution can have harmful effects on the receiver water (fish mortality is the most visible manifestation) as well as on the beneficial uses of water downstream.

A large fraction of contamination is primarily due to suspended matter transported by the flowing water. Many of these particles have a high speed

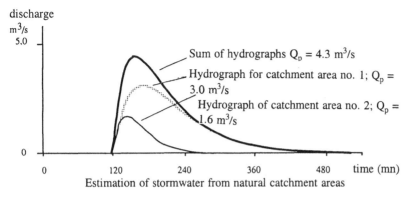

Estimation of stormwater from natural catchment areas

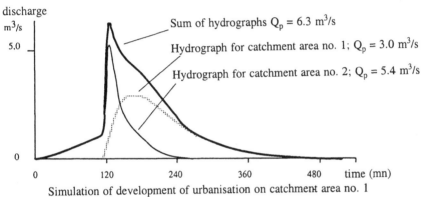

Simulation of development of urbanisation on catchment area no. 1

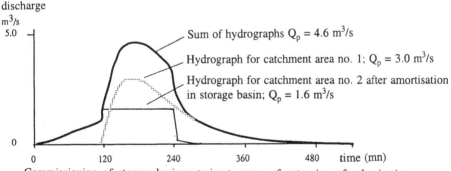

Commissioning of storage basin exterior to zone of extension of urbanisation. Outflow from storage basin equal to maximum discharge of basin in natural state. **Example of utilisation of software for designing sanitary works**. *Note:* For the maximum flow rate in the drainage channel to be identical to that before urbanisation, outflow from the storage basin should be less than that of the natural basin (for the storm event under consideration).

Fig. I.6 Example of computation of a retention basin by the discharge method using the software TERESA.

of settling, which is favourable to good *decantation*. It has long been realised that retention basins have a positive impact on the water quality downstream although originally they were installed merely for protection against floods.

In the context of development of regulations, utilisation of the ability of these works for remediation has become an obsession with designers. Consideration of remediation introduces some constraints related to:

— consideration of objectives regarding the quality of the receiver water,
— protection of the retention basin for planned beneficial usages.

Various possible designs exist for attaining these objectives:

— choice of type of basin,
— inclusion of additional works (pre- or post-treatment).

For the objective of protection of the quality of given receiver water, dimensioning of the basins for effective settlement requires knowledge of the variations in load of pollutants carried by stormwater over a sufficiently long period, generally a year, as well as evaluation of the rate of settlement of suspended matter. Measurement of these is time consuming and laborious, and thus justifiable only for large works. **Furthermore, one must often refer to results obtained on experimental catchment areas and use rather simplified methods of dimensioning.**

I-4.3.1 Characteristics of stormwater pollutants

POLLUTANTS LOADS OF STORMWATER

Tables I.2 and I.3 synthesise the data of average concentrations and annual loads of various pollutants compiled during field studies on experimental catchment areas in France (stormwater of discrete storm drainage networks).

Table I.2 Mean concentration C (in mg/l) and specific annual load M (in kg/ha/yr) in stormwater

Basin drained	Parameter	Pollutant				
		SS	COD	BOD$_5$	TKN	Lead (Pb)
Maurepas (78 ha)	C	190	77	12	3.3	0.085
	M	940	380	55	16.0	0.41
North Ulis (91 ha)	C	440	190	34	6.1	0.12
	M	1100	460	85	17.0	0.30
Aix ZUP (13 ha)	C	300	200	38	5.4	0.16
	M	630	430	75	12.0	0.35
Velizy (78 ha)	C	190	90	17	3.8	0.47
	M	400	190	36	8.0	1.0
Nice (06 ha)	C	130	120	28	—	—
Barron de Berre	M	540	530	—	—	—

DISTRIBUTION OF POLLUTION DURING A STORM EVENT

The questions of whether the first flush of stormwater is more polluted than successive ones and which portion of the stormwater reaching the basin

Table I.3 Annual pollution of stormwater [62]

Parameters	BOD$_5$	COD	SS	Hydro-carbons	Lead (Pb)
Mean concentration (mg/l)	26	179	234	5.3	0.34
Specific polluting load (kg/impervious-ha/yr)	90	632	665	17.0	1.1

should be treated, are presently subjects of intense discussion. Measurements carried out for the total quantity of SS discharged in the Maurepas basin are presented in Fig. I.7. Measurements carried out on other catchment areas yield similar results. Globally, during a storm event, 50% of the pollution is transported in the first 30 to 40% of the stormwater.

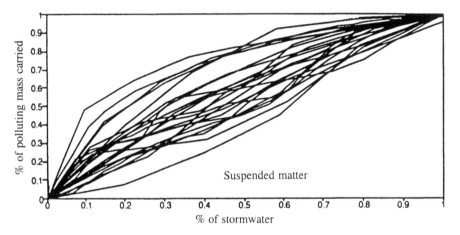

Fig. I.7   Distribution of pollution during a rainstorm event (after Bachoc and Chebbo [1]).

CHARACTERISTICS OF POLLUTANTS IN STORMWATER
Pollutants in stormwater are distinguished by certain features which favour treatment. A large portion of the pollutants attaches to solid materials; only nitrites, nitrates and phosphates are essentially in dissolved form. Table I.4 illustrates this feature for some parameters.

Table I.4   Pollutants attached to settling particles as percentage of total pollution [1]

BOD$_5$	COD	SS	Hydrocarbon	Lead
83 to 92	83 to 95	48 to 82	82 to 99	79 to 99

Granulometry and density: Fine particles constitute the major portion of suspended solid matter. Their density is generally higher than 2.2 g/cm^3.

Settling velocity: Particles tend to agglomerate and their actual velocity of fall is larger than their theoretical velocity (Table I.7). V$_{20}$ represents the

Table I.5 Distribution (in %) of pollutants present in particles of different size ranges (Béquigneaux [33]) for rainfall on 19.10.89 (after Bachoc and Chebbo [1])

Granulometric fraction	COD	BOD$_5$	TKN	Lead
> 250 µm	30	25	13	41
50–250 µm	0	9	3	0
< 50 µm	70	66	84	59

Table I.6 Particle size distribution in stormwater. D$_{50}$ is the median diameter of particles, while D$_{10}$ and D$_{90}$ represent fictitious mesh sizes through which 10 and 90% of the solids would pass (after Bachoc and Chebbo [1])

Type of network		D$_{10}$ (µm)	D$_{50}$ (µm)	D$_{90}$ (µm)	% < 100 µm
Discrete	Mean	7.4	32	617	81
	Standard deviation	1.1	3.5	442	3.3
Combined	Mean	6.8	34	331	75
	Standard deviation	3.3	6.4	112	5.5

Table I.7 Actual settling velocities in m/h for suspended matter in stormwater [1]

Range	V$_{20}$	V$_{50}$	V$_{80}$
Settling velocity	0.7 – 2.4	5.5 – 9.0	22 – 55

settling velocity such that 20% of the suspended matter settles slower than the indicated velocity range.

As shown in Table I.8, this property helps to obtain a relatively large reduction in pollution after only a few hours of sedimentation.

Table I.8 Reduction in pollutants by settling as percentage of total pollution [1]

SS	COD	BOD$_5$	TKN	Hydrocarbon	Pb
80 to 90	60 to 90	75 to 90	40 to 70	35 to 90	65 to 80

Currently available data show that **if obtention of satisfactory remediation by settling alone is desired, particles smaller in size than 50 (µm have to be intercepted.** Simple sedimentation assumes retention of particles whose velocity of fall is less than one metre per hour. At present there are no experimental data providing knowledge of the distribution of pollutants in terms of velocity of particle fall; one has to depend on indirect knowledge of the phenomena.

I-4.3.2 *Possible alternatives for remediation*

As seen above, the characteristics of pollution in stormwater imply that sedimentation is the generally preferred mechanism of treatment. It plays a role in numerous pretreatment systems and even in the basins themselves. Theoretical aspects of sedimentation are discussed in Sec. II-2.2.

When a retention basin has only a strictly hydraulic role, duration of settlement should be as short as possible. Nonetheless, there is a certain compatibility between hydraulics and the process of remediation which can often be usefully employed for improving the water quality.

REMEDIATION IN THE BASIN ITSELF

De facto sedimentation takes place in both dry basins and detention or wet ponds. The latter also serve as the seat of biological processes, which imply reduction of BOD_5 and COD, consumption of nitrogen and phosphorus and increment in dissolved oxygen through photosynthesis as well as atmospheric exchange. Improvement in water quality observed in basins in the Parisian region (Marne-la Vallée, Melun-Sénart) is shown in Table I.9.

Table I.9   Wet ponds: Comparison between water entering from strictly discrete stormwater networks (average values from the literature) and water at outlet (observed concentrations in mg/l for wet ponds) (ecological behaviour as given by SAUVETERRE [77])

	BOD_5	COD	SS
Inflowing water (intermittent flow)	10 to 100	50 to 600	Variable
Outflowing water (average values)	2 to 10	10 to 50	10 to 30
	TKN	Chlorides	Phosphates
Inflowing water (intermittent flow)	3 to 6		
Outflowing water (average values)	1 to 4	10 to 60	0.05 to 1

SEDIMENTATION IN COMPARTMENTALISED BASINS

The primary objective of compartmentalisation is to limit the operations of basin maintenance and desludging. The basin is generally divided into two or three compartments whose volumes may or may not be equal. When the first compartment is full, surplus water spills over the separating wall which acts as a weir. A scheme of operation is given for the case of Perinot (Annexure: Case Studies).

SEDIMENTATION IN RETENTION TANKS (ON A BY-PASS USED IN TIMES OF NEED)

These units are located on a by-pass parallel to the basin and their function is to catch and store part of the volume, when needed. In such tanks, the emptying is deferred during the phase of filling there is no outflow. Removal efficiency depends on the duration of detention of water in the tank and its depth. The optimal depth is generally 2–4 m in the case of emptying by gravity. Table I.10 gives the theoretical orders of magnitude of the dimensions and operation of these systems. The detention times shown in the Table are exclusive of time of filling and evacuation.

Evacuation of water is started as soon as sedimentation is completed. Removal of sludge is carried out as per the methods described in Section III-2.2.3.

Using the results of a series of measurements over long duration on four experimental strictly separate storm drainage systems in catchments of

Table I.10 **Theoretical** detention times in terms of depth of settling, settling velocity and performance

Depth of settling (m)	Settling velocity, V_s (m/h)	Removal of SS (in %)	Minimum detention time (h)
	0.5	85	4.0
2	1.0	80	2.0
	4.0	60	0.5
	0.5	85	8.0
4	1.0	80	4.0
	4.0	60	1.0

surface area extending to several tens of hectares, G. Chebbo determined the efficiency of interception of diverse capacities of stocking storage in terms of various criteria.

Table I.11 presents the percentage of SS intercepted in terms of the volume of settling tank per impervious hectare. It shows that settling tank volumes of 100 to 200 m³ per impervious hectare are necessary for intercepting a significant portion of the pollution. Efficiency is given by the product of the percentage of intercepted SS and the removal efficiency which depends on time of settling, as shown above.

Table I.11   Percentage of SS intercepted as a function of volume stored per impervious hectare of catchment [16]

Volume of settling tanks (m³/imp. ha of drainage area)	%age of mass M of critical particles removed annually	Intercepted %age of mass during storm events	Number of events with very low removals (no./yr)	
			Average events between M_x 1% and M_x 5%	Big events $M_x > 5\%$
20	36 – 56	5 – 10	4 – 14	2 – 4
50	57 – 77	13 – 29	2 – 10	1 – 3
100	74 – 92	26 – 74	2 – 4	1 – 2
200	88 – 100	68 – 100	1 – 3	0 – 1

REMEDIATION ASSEMBLIES

Retention basins, often accompanied by treatment works, are discussed in Section II-2.2. Some samples of simple remediation arrangements are presented in Tables I.12 and I.13.

I-4.3.3 Software for water quality

The earliest softwares for water quality were developed in the 1970s. Since then little progress has been made in mathematical models, undoubtedly because of the great complexity of the phenomena involved in generation of pollution as well as its entrainment and transport in sewer networks, and also the difficulty of evaluation of impacts on the environment.

Table I.12 Possible options for remediation of stormwater in dry basins

Examples of remediation options in dry basins	Advantages	Drawbacks
 Compartmentalised dry basin	■ Reduction in cost of maintenance, especially if floor of first compartment is made of concrete	■ Little reduction of impact downstream: no additional dilution no additional self-purification in basin
 Pretreatment plant / Dry basin	■ Reduction in cost of basin maintenance ■ Interception of accidental pollution ■ Generally more favourable to remediation ■ **Observation**: Compartmentalisation still possible	■ Same ■ Pretreatment work has also to be maintained
 Dry basin / Lamellar separator	■ Improvement in final treatment with regulated flow ■ Interception of accidental pollution ■ **Observations**: Dimensioning of separator should take into account the decantation obtained in basin. Layout generally chosen for retention of hydrocarbons. Efficiency of removal of SS controversial ■ Compartmentalisation possible	■ Same

Table I.13   Possible options for the treatment of stormwater in detention ponds

Examples of remediation options	Advantages	Drawbacks
Pretreatment work	■ Protection of basin against accidental pollution	■ Pretreatment work has also to be maintained
Detention pond / Lamellar separator	■ Improved final treatment with regulated flow ■ Possibility of destruction and/or confinement of accidental pollution ■ **Observations:** Dimensioning of separator should take into account sedimentation in basin	■ No protection of basin against deposition of accidental pollution ■ Lamellar separator has also to be maintained
Dry or wet compartment / Detention pond	■ Layout generally unchanged for retention of hydrocarbons. Efficiency of SS removal controversial ■ Reduction in cost of basin maintenance	

Nevertheless, aided by calibration with series of in-situ measurements, the softwares do succeed in predicting, albeit imperfectly, the flow of pollution into the receiver waters. The major softwares developed in France are:
— FLUPOL, for simulation of inflow flux of pollutants into the sanitary networks; KAPLAN, for the quality of water in transit to rivers. These two softwares were developed by 'Agence de l'eau Seine Namandie' in collaboration with 'Syndicat des Eaux d'Ile de France' and 'Compagnie Générale des Eaux'.
— CONVEC, developed by SOGREAH.
Mention may also be made of foreign softwares such as STORM, SWMM, MOSQUITO, MIKE 11 and SAMBA, HYDROWORKS, MOUSE etc.

I-4.4 Case of Infiltration Basins

I-4.4.1 Introduction

Infiltration basins are generally installed when an outlet is not available and/or when permeability of the soil is good. Evacuation takes place by infiltration across the floor and the sides and may be facilitated or accelerated by providing percolating ditches. These basins are generally constructed by digging out to create depression and thereby precluding risk of rupture of dyke.

Infiltration of water in the soil may have harmful effects such as dissolution of rocks and rise in water table, which can cause flooding of the subsoil or damage the foundations and walls of adjacent buildings through rise of capillary water.

This type of basin should be proscribed if the pollution carried by stormwater is apt to threaten the quality of an aquifer with potential for exploitation and a fortiori within the perimeter of the protected area.

I-4.4.2 Computation of infiltration

In soil hydraulics simple methods of computation are based essentially on Darcy's law. It should be verified whether the case under consideration satisfies the conditions of this law. Some global and simple formulas applicable to typical situations frequently encountered are given here (see Table I.15). More complex cases are tackled by more elaborate methods such as analysis of flow networks by electrical analogy or numerical resolution. Models exist for computation of flows in transition regimes using other laws of infiltration.

WATER TABLE AT GREAT DEPTH AND TERRAIN HOMOGENEOUS
Flow is vertical, following a gradient close to 1, and varies very little with height of water in the basin. Rate of infiltration is given by application of Darcy's law:

$$Q \, (m^3/s) = K_v \times S$$

where S is the surface area of infiltration (equal to surface area of the water body) and K_v is the coefficient of permeability in m/s. Table I.14, constructed from the above relation, gives some idea of infiltration loss in terms of the coefficient of vertical permeability.

Table I.14 Infiltration rates and time for emptying of a water body as a function of the permeability of the bed material

Nature of terrain	Vertical permeability (m/s)	Rate of infiltration (m³/day/ha)	Time for complete evacuation of water 1.5 m deep
Clay	10^{-9}	0.86	> 45 years
Marl	10^{-8}	8.64	> 45 months
	10^{-7}	86.40	< 6 months
Silt			
Fine sand	10^{-6}	864	> 20 days
	10^{-5}	8640	> 2 days
Coarse sand			
Fissured rock	10^{-4}	86,400	> 4 hours
	10^{-3}	864,000	< 20 min

When the terrain separating the bottom of the basin from the water table is composed of n different layers of thickness (a_i) and vertical permeability ($K_{v(i)}$), the equivalent permeability (K_v) is obtained from the following relation:

$$K_v = \frac{\sum\limits_{1}^{n} a_i}{\sum\limits_{1}^{n} \dfrac{a_i}{(k_v)}}$$

When the water table is at great depth and the terrain bilayered with $K_2 \gg K_1$, the lower layer 2 is considered a drain for the upper layer 1. In this case, the rate of head loss depends entirely on the upper layer of smaller permeability K_1 and thickness a_1.

BASIN IS TANGENT TO OR CUTS WATER TABLE IN HOMOGENEOUS MEDIUM
Schneebeli's relation developed for drilling in a deep permeable layer may be used in an injection mode for such cases (see Table I.15).

BASIN ENTIRELY TRANSVERSE TO AQUIFER
Infiltration formulas are rarely used in this case. The standard Dupuit formulas may be resorted to if the flow can be considered steady and with radius of influence fixed in time.

Table I.15   Simple formulae giving rates of infiltration in common cases

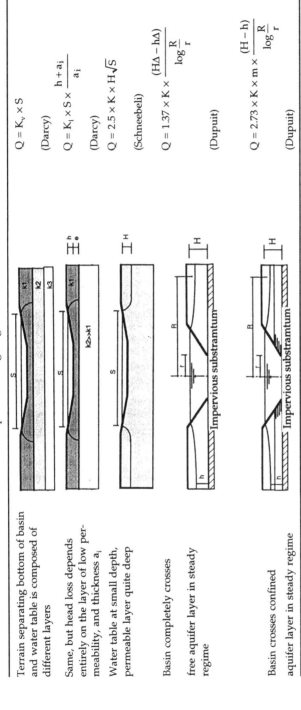

Terrain separating bottom of basin and water table is composed of different layers	$Q = K_v \times S$ (Darcy)
Same, but head loss depends entirely on the layer of low permeability, and thickness a_i	$Q = K_l \times S \times \dfrac{h + a_i}{a_i}$ (Darcy)
Water table at small depth, permeable layer quite deep	$Q = 2.5 \times K \times H \sqrt{S}$ (Schneebeli)
Basin completely crosses free aquifer layer in steady regime	$Q = 1.37 \times K \times \dfrac{(H\Delta - h\Delta)}{\log \dfrac{R}{r}}$ (Dupuit)
Basin crosses confined aquifer layer in steady regime	$Q = 2.73 \times K \times m \times \dfrac{(H - h)}{\log \dfrac{R}{r}}$ (Dupuit)

SOIL CLOGGING

Clogging can very rapidly endanger the operation of an infiltration basin, more so if the characteristics of permeability and structure of soil are not very favourable. Also, clogging must be taken into account right at the stage of preliminary planning. To limit or manage soil clogging, one may:

— reduce the silt content in the inflow or introduce decantation ahead of the basin,

— anticipate the need for intervention when clogging occurs, by placing a layer of sand on the bottom, which will play the role of a periodically renewable filter.

I-5   POTENTIALS AND CONSTRAINTS IN PLANNING

I-5.1   Land-use

In urban zones, the cost of land and the stipulations by decision makers that no space be left idle, as well as the demands of people for improving environment and quality of life, are factors that force developers to provide retention basins for leisure and recreation.

I-5.1.1   Potentials of development

For a given frequency of failure the size of retention basin needed is directly proportional to the surface area of the related catchment; however, the size of facilities for recreation depends essentially on the intensity of urbanisation and the type of recreational facilities involved, as shown in Table I.16.

Certain facilities, such as play grounds, etc. may necessitate fencing the area and require caretakers for surveillance. User safety is thereby strengthened even though the risk of accidents due to storm or flooding is very minimal given the rather slow rise of water in the basin and the low frequency of such flooding.

Other facilities, such as a golf courses, gymnasia, sports fields etc., take up considerable space and thus are particularly suitable for installation of basins.

A good correspondence is often observed between the space needed for installation of retention basins and that needed for development of open air and sports activities.

With due consideration of the constraints related to hydraulics and/or remediation mentioned above, retention basins adapt to most of the requirements resulting from ancillary usages.

I.5.1.2   Constraints to development planning

Once it has been decided in principle that a multipurpose retention basin is to be installed, possibilities for integrating various usages of the basin are

Table I.16   Classification of open air and green recreational spaces (from DE SABLET [22])

Type of location	Type of facility	Surface area of facility	Space management	Radius of influence
Scattered housing (200 – 500 family units)	Playground for children	500 – 1000 m²	Private or public gardens	30 – 100 m
	Area for adult relaxation			
Residential neighbourhood (1000-2000 family units)	Playground for children (super-vised or other-wise)	2000 – 5000 m²	Public parks	
	Game fields, gardens, parks PT gymnasia for students	1000 – 3000 m² or more	Sports admini-stration	250 – 300 m
	General gymnasia Walkways	800 – 1500 m²	Local administration	
Residential townships (4000 – 5000 family units)	Small parks	300 – 500 m²	Gardens, parks around buildings	
	Large parks	6 – 10 ha		
	Playgrounds	5 ha or more	Public places maintained by	500 – 800 m
	Leisure centres	1.5 – 3 ha	local admini-stration	
	Sports grounds Walkways			
City	Multipurpose sports stadia	10 ha or more administration	Central urban administration	800-2000 m
	Urban parks	20 – 30 ha		
Suburbs	Roads and tracks for automobiles	4000 – 5000 m²	Exclusive zones	2000 m or more
	Open air and leisure areas	100 ha or more	Local green areas administration	
	Golf courses	20-50 ha		

determined by multiple constraints related to acquisition of land, hydraulic aspects, requirements for areas of leisure and recreation as well as consider-ations of various likely conflicts of usages.

CONSTRAINTS RELATED TO OCCUPATION OF LAND
Conditions for installation of retention basins differ markedly according to the site—whether it is virgin land or a developed area. In the case of virgin land, retention basins are generally welcome: they are installed before the arrival of inhabitants and though not considered in the highly developed category, they do increase land value and quality of life in the area. In such cases, the basin planners generally have considerable latitude in designing their project; the only person they have to satisfy is the chief of works.

The situation is radically different for developed sites. Habits are often well entrenched and the chosen site may be under use for some unrestricted activities (promenade for domestic animals) or a place of quietude for the inhabitants. Acquisition of land under green areas poses a problem since their area of influence extends beyond their immediate premises. Understandably installation of a retention basin may meet very firm resistance from resident social groups, which invoke such arguments as: 'upstream urbanisation creates nuisances and hence installation of a retention basin here is not justifiable', or 'the basin will generate foul odours', or 'risks of children drowning are high', or 'basins perturb traffic and have an adverse effect on commerce and business' and so forth.

Planning a retention basin should hence be based on an in-depth diagnosis of existing urban activities and attitudes.

CONSTRAINTS TO PLANNING AREAS OF LEISURE, OPEN SPACES AND
SPORTS GROUNDS

Every construction has its contraints of development and operation. The general factors and considerations relevant to planning urban public spaces (DE SABLET [22]) are listed below:
— nature of surrounding constructions; collective housing, dispersed or individual houses, industry, sports areas, schools etc.;
— number of users and features of utilisation;
— nature of the site and climatic conditions;
— interrelationships of the selected area with the surrounding spaces and structures, which affect various aspects of life of the residents, namely commerce, leisure, services, domestic animals etc.

CONSTRAINTS OF HYDRAULIC PLANNING

These constraints have already been detailed. They pertain to the topography of the site, depth of water table, shape of the basin, location of works of inflow and outflow, hydraulic operation, nuisance potential, nature of materials used and vegetation developed.

COMPATIBILITY OF PLANNING

This is generally achieved by one or/and other of the following:
— expanding the size of the works. This may pose difficulties if the project is studied after the land use plan (POS*) has been approved, or, for smaller projects, after the master plan has already been finalised.
— reduction of expectations from proposed works and satisfaction of needs on other sites;

---

*'Plan d'Occupation des Sols' which is a public document specific to each city dealing with urban planning.

— reduction of frequency of failure of the basin and/or development of complementary hydraulics on the upstream areas so as to reduce the inflow or/and on the downstream areas so as to increase the transport capacity of the collectors.

I-5.1.3   *Conflict among various usages*

These apparently arise primarily from:

— lack of public information on the initial hydraulic operation of the works. It should be ensured that information relevant to municipal regulations, instructions for public associations and the public in general, giving the modus operandi and maintenance requirements for the project, are widely disseminated. This information may be displayed in the form of notices placed near the main access, as shown in Photo 3.

— filling up too often or further nuisances due to a lack of maintenance of the stormwater network (bad connection for example). The first point is generally related to an under-estimation of the amount of precipitation due to a higher impermeability of upstream basins in comparison to that taken into account at the project stage.

Photo 3   Typical notice of information concerning a wet pond.

— too many visitors on the site in connection with initial planning. This is particularly relevant to the case of wet ponds for which the pollution generated by the activities can degrade the quality of water in the basin (see MALET and TA [50]).

I-5.2 Ecology

A priori it appears difficult to formulate strict rules regarding the shape and dimensions of basins for storing stormwater which would ensure satisfactory operation from the ecological point of view, for the simple reason that this concerns a living environment. The parameters that govern the activities of such an environment are just too numerous. In discussing this subject it would be useful to first note some of the features of natural water bodies.

I-5.2.1 Natural waterbodies

There are two major categories of natural water bodies:
— Lakes of moderate to large depth (5–6 m or more), sufficient for setting up a thermal stratification from top to bottom and a movement or displacement of a thermocline during the course of a year. The ecology of these water bodies is very specific, given the presence of this thermocline with strong possibilities of degradation of the water quality (non-aeration of the water at depth and anaerobic phenomena at the bottom).
— Ponds, in which the depth does not suffice for the establishment of thermal stratification.

Wet ponds resemble ponds because they are usually shallow and the type of water and immediate environment (banks, vegetation etc.) are similar.

Essential knowledge with respect to ecological behaviour of water bodies is provided by observations made in limnology [30]. In particular, a water body during the course of a complete cycle of seasons will, so to speak, discharge or decompose the biomass it produces: obviously, a suitable mineralisation of organic matter is needed for this purpose.

In our present state of knowledge, there are no rules to assist in computing designs for ecological equilibrium for a water body with landscaped banks. One can only propose some principles which favour creation of a good-quality aquatic environement.

The concerned objectives include:
— Water
   * absence of foul odours,
   * transparency: 60 cm to 1 m,
   * colour: not very marked,
   * quality: favourable for the life and reproduction of fish;

— Vegetation on and around the water body, say on the banks: easy to control, exhibiting good landscape effect;
— Maintenance: easy and enabling conservation of a pleasing appearance.

I-5.2.2   Ecological considerations in basin dimensioning

Bank length plays a more important role than simply selection of a general shape for ecologically harmonious development. Curved banks are preferable to rectilinear ones because they provide increased length of banks.

The waterward slope of embankments should be gentle to enable planting different varieties of plants. It is desirable to create groves of trees to facilitate fish movement and creation of 'fish nurseries'.

The mean depth of a water body should be of the order of 2 m. High bottoms or islands favour vegetation. Deepwater pockets (3 m or more) are a source of fresh and oxygenated water for fish. It is therefore preferable that the bottom relief be irregular.

As they are generally not accessible to the public, islands offer shelter to flora and fauna in addition to that provided by the banks.

Plants, bushes and trees are helpful in preventing damage during brief periods of submersion or when the basins are being filled or evacuated. Nevertheless, a marling of about 50 cm (see I-6.2) is generally provided for a 10-year rain cycle.

I-5.2.3   Polluting loads

The smaller the load of pollution received by a water body, the better its behaviour. Several reports have noted the ravages caused by excessive pollution:

— **Waste water discharges during the dry-weather periods:** even a small number of improper connections of waste water with a storm drainage network can lead to strong bacteriological pollution and accelerated eutrophication.
— Large quantity of stormwater flow over dense urban facilities such as roads, parking lots: specific treatment of inflow is needed.

I-5.2.4   Maintenance of water level

A low water level (at the end of a dry season) often conditions the ecological equilibrium of a basin. Deficiency of water may cause excessive heating of the water body, algal proliferation accompanied by 'water flowers', (bloom), excessive weed growth etc.

I-5.2.5   Diversity of biotopes

Diversity of biotopes implies symbiosis between species. Interspecific concurrence facilitates multiplication of food chains and ensures·population self-control. For this to happen, however, the water depth, contour of banks, slope of embankments, exposure to sun, nature of the bottom etc. all have to be planned. Subsequently, useful species of vegetation and animals are introduced to complete the panorama and natural colonisation.

I-5.2.6 *Vegetation of the basin and banks*

Vegetation, an element of the biotope (it constitutes a habitat for diversified fauna) and biocoenosis (comprising living organisms, producers of biomass) deserve special attention at the design stage of basins. First, the case of detention ponds in which three types of vegetation may be present shall be discussed and then the case of dry basins involving numerous rules of design for parks and gardens and, to some extent, knowledge of wetlands.

CASE OF DETENTION PONDS

Before examining the types of vegetation present in and around such a water body, it is important to remember that vegetation plays several important roles:

First of all, it is considered an important element of beautification. This effect is best achieved by a variety of plants around the basin, providing change of colours with the seasons, and carpets of waterlilies and other aquatic plants on the waterbody.

Strengthening role. Vegetation is an important and inexpensive element of bank stabilisation and protection against erosion; it also provides shelter for fauna.

Source of nourishment. Vegetation produces biomass, the starting point of food chains. Many animals are herbivores: fish, birds and gastropods. Innumerable insects, amphibians, reptiles and mammals depend for their existence on this biomass and those who consume it.

Large capacity of water purification. **Vegetation participates in the consummation of nutrients**. Nitrates and phosphates are widely consumed by the roots of higher plants and by plankton.

Upon death, plant tissues decompose in water; a large portion of these elements is mineralised and can be released in winter, the season in which the flow of the receptors can best accept it. A small portion settles and forms benthic silt. **The rate of silt deposition in a well-functioning water body is a few millimetres per year.** But it must not be forgotten that should biomass production be higher or its evacuation insufficient, dysfunctioning of the basin will rapidly set in, given the fact that the layer of benthic silt can attain a thickness of several centimetres in one season.

Aquatic vegetation can act as a **natural agent of purification by assimilating certain dissolved metals and even some organic compounds**; for example, phenols are assimilated by water-hyacinth. However, this is more a matter of storage in the plant, which can only disappear by exportation.

Immersed stems and leaves serve as support for numerous bacterial colonies which decompose the organic matter contained in the water.

Vegetation **regulates the development of phytoplankton.** Thanks to this concurrence as well as the better development of populations of herbivorous fish which consume this plankton, limpid water is maintained.

Diverse forms of aquatic vegetation. At the stage of designing a basin for stormwater for which one tries to make it look like a natural pond, a good equilibrium should be found among diverse forms of vegetation. It is best to study the spontaneous vegetation of nearby ponds, which will be analogous in climatic and edaphic features. The biocoenoses should be maximally diverse and the maximum number of plant and animal species encouraged in order to initiate interspecific concurrence.

In some basins whose bottom and walls are cemented, generally monospecific vegetation (most often waterlilies) is planted in immersed beds. The growth of such plants must be strictly controlled, however.

It is further worth noting that vegetation grows in concentric belts in water bodies. In proceeding from the bottom of a basin or pond towards the banks, one finds:

— **Hydrophytes** (Fig. I.8) rooted in the benthic silt whose foliar system is totally submerged (*Elodea, Potamogeton, Ceratophyllum* etc.) or the leaves spread on the surface (waterlilies). Duckweeds are free hydrophytes. Most of these weeds flower on the surface. Some of the hydrophytes commonly found in stormwater basins are: *Ceratophyllum* sp., *Lemna polyrrhiza minor, Myriophyllum* sp., *Nymphea alba, Polygonum amphibium, Potamogeton cripus, P. lucens, P. natans* and *Ranonculus divaricatus.*

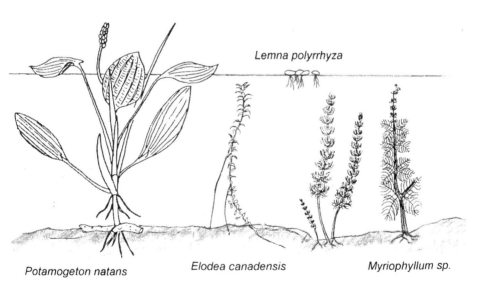

Fig. I.8 Some hydrophytes commonly found in stormwater basins.

— **Helophytes** (Figs. I.9 and I.10), also rooted in the benthic silt but with the greater part of their leaves aerial; these include some groups of herbaceous vegetation along the banks.

Some helophytes commonly found near and on banks of stormwater basins are: *Alisma plantago, Alopecurus geniculatus, Carex* sp., *Epilobium* sp., *Iris pseudacorus, Juncus* sp., *Lycopus europaeus, Mentha aquatica, Myosotis palustris, Phalaris arundinacea* and *Sparganum* sp.

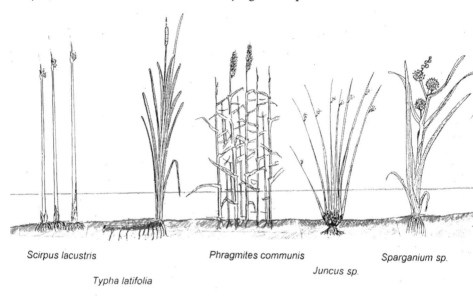

Scirpus lacustris

Typha latifolia

Phragmites communis

Juncus sp.

Sparganium sp.

Fig. I.9 Some helophytes commonly found in stormwater basins.

— **Angiospores** and woody vegetation growing on the banks, consisting of trees which can withstand a highly humid environment (willows, alders, poplars etc.). Mention should also be made of the exotic fragrance of bald cypress and of pneumatophores which provide remarkable protection against vermin.

The choice of plant species to be sown or zones where vegetation is expected to grow spontaneously, should be made right at the stage of design of the contours of the bottom of the basin and construction of banks.

Acclimatization of imported (or exotic) vegetation is rather risky. Some undesirable proliferations originate from an imported plant which, on not adapting to coexistence or not being consumed, grow in a disorderly fashion. A plant such as water-hyacinth, well known in tropical regions, is a classic example of vegetation to be avoided (except when biomass for fuel or other uses is expressly desired). In French climates, introduction of 'jussie' and elodea created an ecological imbalance. Similar precautions have to be taken in regard to animal species.

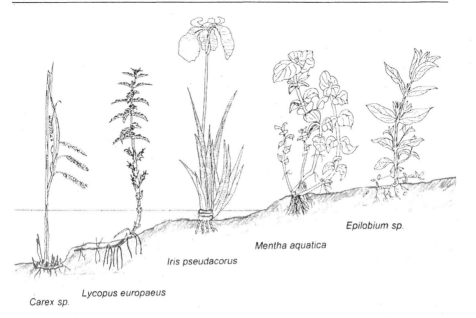

Epilobium sp.

Mentha aquatica

Iris pseudacorus

Lycopus europaeus

Carex sp.

Fig. I.10   Some helophytes found on banks.

CASE OF DRY BASINS

Vegetation of dry basins, natural or introduced, is the same as that usually found in parks and gardens. The rules of design are the same and take into account usage of the basin beyond storm episodes.

Distribution of plant species depends on the water table level under the bottom of the basin; if it touches the surface even in the dry season, species of trees loving humidity will be present: willows, alders, poplars etc.

One may also observe spontaneous variations in the distribution of plant species depending on possible duration of submersion, changes in soil and eventually the microclimate at the bottom of the basin (e.g., possible freezing in case of deep basins with steep embankments). The plan for greening must take into account all the hydraulic, climatic and edaphic aspects.

I-6   SHAPE AND MORPHOLOGY OF RETENTION BASINS

I-6.1   General Shape

The shape and general contour of stormwater detention ponds are decided to a great extent by local topography, land parcelling and planned functions and uses. In plan the shsape may be quite varied but the lengthwise profile generally presents the following zones as one proceeds from upstream to downstream:

— upstream zones of expansion (flooding zones) form the tail of the basin;
— outfall of collector(s), with or without a device for grit removal or a pretreatment facility;
— basin itself, may or may not be dotted with islands;
— crucible-shaped portion enabling emptying of the basin, collection of fish, and removal of sediments and silt;
— calibrated overflow weirs at nominal water level or at FRL and flow-control system;
— dyke (if there is one).

These elements are schematically depicted in Fig. I.11.

Flooding zone

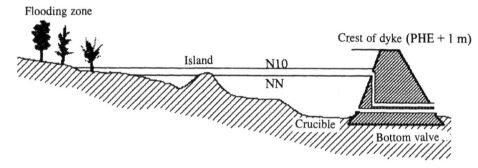

Fig. I.11   Lengthwise profile of a detention pond.

I-6.1.1   Detention ponds

Banks with varied contours and irregular slopes offer a more pleasing view and better diversification of biotopes.

The geometrical shapes like those of sand pits or stone-quarry drives, for a given surface area, having a minimum length of banks. They should be avoided in particular for fish ponds and basins with aquatic gardens.

Geometrical forms may be locally adapted in an existing urban fabric or when dictated by the architectural environment. Treatment of banks, lined or unlined, helps in upgrading the major portion of these rectilinear configurations. A rectilinear shape may also be dictated if the basin is used for some type of water sports. Basins in the form of rectilinear or curvilinear canals also have an aesthetic aspect. Numerous examples of such basins exist, especially in the Netherlands, where they constitute the most extensive form of storing stormwater.

I-6.1.2   Dry basins

A circular shape affords the minimum length of embankment and involves minimal cost in earthwork and maintenance. In some cases the shape determines the ancillary uses of the basin beyond the rainy season, for example, basins used as stadia, hypodromes, velodromes etc. Constraints of maintenance can also be an influencing factor. Safety considerations likewise dictate provision of exit routes in case of rapid water rise.

In particular, isolated mounds are proscribed in dry basins.

I-6.1.3 *Underground storage basins*

The shape of underground storage basins is determined by the land-uses and the technical options available: it may be parallelepiped, cylindrical, prismatic etc. The wall shape may be affected by geotechnical constraints and distribution of load due to constructions.

I-6.2 Useful and Total Depth of Detention Ponds

The total depth of a permanent water body depends on the initial topographic conditions and planned usages.

The useful depth (or marling) is the variation or difference between the nominal water level outside the flooding season and the level attained during rains when the full volume of a given flooding frequency has been stored in the basin.

Marling of the order of 50 to 70 centimetres for a ten-year cycle, along with a mean water depth around 2 metres allows:

— effective dilution of the streaming flow by the quasi-steady mass of water standing in the basin before flood flows arrive. No significant variation in planktonic population occurs. On the contrary, if the volume of stormwater is much larger than the initial volume in the basin, a complete renewal of water mass takes place and aquatic biocoenoses, especially of planktonic forms, have to be re-established. Shock flooding has very adverse effects on the ensemble of water biota;

— easy treatment of banks (lined and unlined) at low costs;

— imperceptible variation in water levels during frequent light showers, thereby preserving the aesthetics.

A marling of 50 centimetres for a ten-year cycle almost corresponds to a marling of 100 to 120 centimetres for a 100-year cycle. A topographic layout of basins, wedged at a sufficiently low level in relation to the surroundings so that the collectors can flow under gravity, allows storage of stormwater with no inconvenience other than a slightly larger flow load on the relevant drainage channels. Inadequacy of such storage would be rather unusual.

A low marling value also helps in tackling the problem posed by a slight increase in urbanisation in the catchment area; a surcharge of 10% yields an additional elevation of about 5 centimetres in water level for a 10-year flood, which is hardly perceptible. Thus it provides greater security vis à vis urban devisification (which occurs quite frequently) and also vis à vis exceptional rainfalls.

Around Paris, it has been demonstrated that a once-in-10-year flooding needs a marling of 50 cm and the surface of the water body needed for

storing the stormwater represents about 4 to 5% of the strictly built-up and impermeabilised surface.

I-6.3   Approaches

Several types of approaches should be considered while designing a stormwater retention basin.

I-6.3.1   *Approaches for maintenance and safety*

These pertain essentially to approaches to the various stormwater inlet works networks and control systems, remediation units and also supply points of water for fire-fighting. The accesses, constructed above the highest level of water in the basin, should be usable at all times. The corresponding paths should be capable of bearing the weight of vehicles. These approaches shall also include ramps for water intake and rescue boats as well as maintenance of the water body.

Approaches to the works and structures of an underground storage basin should be defined with great precision: access for maintenance of screen, pumps, outlet works, sludge outlets etc.; all such approaches should be co-ordinated with urban planning of surface.

I-6.3.2   *Promenades/Walkways*

These paths should be designed with an eye to aesthetics, aquatic as well as terrestrial. They may also accommodate light vehicles for cleaning and maintaining the roads, green spaces and banks.

I-6.3.3   *Other special approaches*

These include approaches used for:
— parking lots,
— fishing spots,
— ramps for launching small boats, sailboards etc.,
— areas for relaxation or picnics along with associated buildings,
— security posts and life-saving posts,
— measurement and monitoring facilities.

These approaches should be suitable for traffic and are generally asphalted, or for aesthetic considerations, metalled, paved etc.

I-6.3.4   *Safety of approaches to open basins*

Pedestrian paths should be gently sloped with steps placed for any steep descents and ascents since ordinary soil or gravelled paths may have slippery spots due to humidity or earlier temporary submersion.

In the immediate vicinity of water, safety arrangements vary with the type of banks. In the case of vertical banks, unless a security network (watchmen, parapets, barriers) has been provided, there must be a horizontal berm at least 1.50 m wide at a small depth below the top of the bank for easy recovery of a child or an adult who may have fallen.

Public information regarding risk of falling into the water and rapid rise of water level must be prominently displayed; notices should preferably be placed at all entrances, path crossings etc.

I-6.4   Banks

I-6.4.1   Detention ponds

The two main types of banks that can be used in alternation, around a retention basin for stormwater are unlined and lined. These banks serve several purposes:
— aesthetic, through geometrical shapes and nature and quality of surfacing material;
— mechanical—they must be resistant to erosion and slippage;
— biological—they serve as an interface between aquatic and terrestrial media and are colonised by a rich and varied flora and fauna;
— social—the inhabitants should be able to visit them for pleasure (promenade, fishing, other usages) without danger.

NATURAL BANKS
These are quite common. Their profile can be highly varied according to the inspiration of the landscape artist and the types of vegetation chosen. Slopes depend on soil resistance to erosion and sliding, i.e., the constitutive material determines the angle. A typical section in a natural bank with protection against erosion is shown in Fig. I.12.

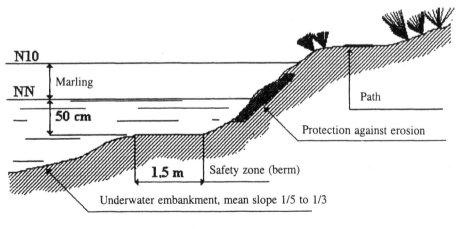

Fig. I.12   Typical profile of an unlined (natural) bank.

Banks exposed to waves have to be reinforced by geotextiles to hold the soil in place until woody plants have taken root; these plants will subsequently ensure bank stability. One may also use honeycomb slabs, stonepitching, bricks etc., especially when overhanging promenades, fishing spots etc. are envisaged. Photo 4 shows a very common type of bank for which the possibilities of aesthetic diversification are unlimited.

LINED OR 'CONSTRUCTED' BANKS

The 'constructed' banks may be vertical or formed by quays, logs, faggots, planks, parapets

Depending on the nature and quantity of materials involved, banks may also be in the form of embankments with more or less steep slopes, protected by rocks, honeycomb slabs, stone-pitching or lining made of concrete. Whichever, the quality of the material is very important, whether it be rot-resistant wood (azobe) or stone-pitching, rocks, slabs or concrete lining (Fig. I.13).

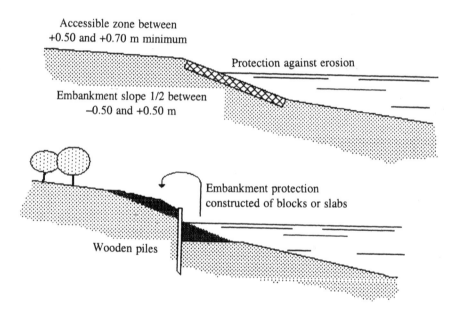

Fig. I.13 Typical profiles of common lined or constructed banks for stormwater basins.

I-6.4.2 *Dry basins*

In the case of dry basins the term embankment is more appropriate than bank. Embankments delimit the most often inundated portions and are governed by the rules stated for detention ponds.

In general, except when the basin is designated for specific sports, embankments are so inclined that one can walk on them to maintain the grass and plant cover. Their slope (height/base) is so designed as to also ensure

Photo 4   Natural bank. By providing fishing spots, deterioration of the banks due to stamping is prevented.

the stability of the constitutive material, irrespective of the stage of plant growth (possible seasonal variations).

I-7   EMBANKMENTS AND DYKES

I-7.1   Basic Requirements

Scientifically speaking, embankments and dykes are similar; the term 'dyke' is used when a waterway is closed and 'embankment' when it is merely confined. 'Constructed' embankments and dykes may be vertical or consist of benches, berms and slopes. Vulnerability to failure increases with embankment height. Thus French regulations stipulate the height and for any embankment exceeding this limit, consultation with the permanent technical committee for dams, constituted by the decree of 13 June 1966 in France, is compulsory. This limit was fixed at 20 metres between the top of the work and the lowest point of the natural terrain in a state of equilibrium.

Other specifications, pertinent to the three conditions listed below, are also mentioned in these regulations; the committee must be consulted regarding these even for embankments of less than 20 metres height:
— height of embankment greater than 10 metres;
— difference between level of the lowest point of the foundation and that of the top more than 20 metres;
— serious consequences feared in case of failure of the public safety works, relevant to the volume of the basin and the proximity of inhabitants living downstream.

It should be noted that the dimensions of most of the existing works constructed in urban areas are smaller than those stipulated and thus consultation with the committee is not necessary. But as there are always people living downstream, the design and follow-up of embankments do demand special care.

I-7.2   Structure of Embankments and Dykes

I-7.2.1   Types and profiles

Depending on the soil or other material available on site and the size of the work to be constructed, there are three major types of dykes and embankments (Fig. I.14):
— **homogeneous** with or without facilities of internal drainage, with or without impervious lining upstream and built with construction material of low permeability (classes A1, A2, A3 of the soil classification by GTR (Guide des Techniques Routières)* [43];

---

*Guide of Road Techniques

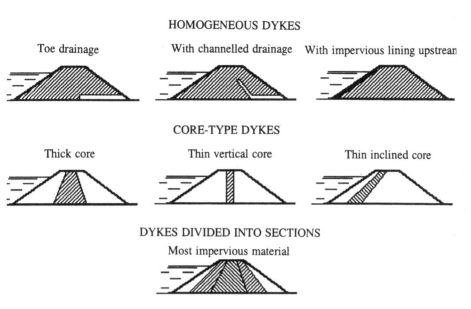

HOMOGENEOUS DYKES

Toe drainage   With channelled drainage   With impervious lining upstream

CORE-TYPE DYKES

Thick core   Thin vertical core   Thin inclined core

DYKES DIVIDED INTO SECTIONS

Most impervious material

Fig. I.14   Common profiles of earthen dykes and dams [41].

— **core type** composed of permeable materials with an internal impervious core, impermeability being ensured by a thin or thick core of materials of low permeability with an associated membrane in some cases. The core is made of soil of low permeability (classes A2, A3, A4 and C1 of the soil classification by GTR) or is constituted as a wall made by an injection of bentonite-cement;

— **dykes divided into sections** characterised by refill of materials of different kinds, the central portion occupied by material of the lowest permeability.

Entirely homogeneous dykes without facility for drainage or impervious lining are used exclusively for works of small size situated in a hydraulically favourable area (deep water table, homogeneous good soil foundations with good drainage characteristics).

I-7.2.2   *Dimensional characteristics*

In addition to the level of the highest water surface, the general profile of a dyke depends on the freeboard, breadth at the top and slope of the sides [54].

FREEBOARD
This is the section between the highest level of the water surface and the top of the dyke. Its function is to ensure protection against any possible overflow over the top of the dyke under the effect of waves or possible extra rise of water level under the action of strong and steady wind. It also provides

safety in case of accidental rise of water in the water body above the highest planned basin water level due to uncertainty in determination of flood flows.

It is prudent to adopt the minimum value of freeboard at 1.20 to 1.50 metres for dykes less than 10 metres in height and at 1.50 to 2.0 metres for works which are 10 to 20 metres in height.

BREADTH AT THE TOP

The breadth at the top of an earthen dyke should be sufficient to ensure that seepage of water across the dyke is not large near its crown when the basin is full. It should also allow movement of equipment required for completion of the work and ultimately its maintenance.

In practice, the breadth at the top of an earthen dyke is never less than 3 metres. For works greater than 9 metres in height, the value of the breadth is often chosen as equal to 1/3rd the height.

SIDE SLOPES

The slope of the side of an embankment or dyke is the ratio of its height to its horizontal projection the ground. It is decided by the conditions of mechanical stability of the fill material and its foundations.

To determine the slope of sides or facings, one generally starts with slopes which appear to be optimal, taking into account the nature of materials, and checks that the dyke is safe with such slopes by studying the stability.

Some values are given in Table I.17 as initial information, which, of course, must be confirmed through a study of stability.

Table I.17: Side slopes of embankments for different conditions

Height of dyke	Type of dyke	Slope of embankments	
		Upstream	Downstream
< 5 m	Homogeneous	1/2.5	1/2
	Built in sections	1/2	1/2
5 to 10 m	Homogeneous, sand fill	1/2	1/2.5
	Homogeneous, with high clay %age	1/2.5	1/2.5
	Built in sections	1/2	1/2.5
10 to 20 m	Homogeneous, sand fill	1/2.5	1/2.5
	Homogeneous, with high clay %age	1/3	1/2.5
	Built in sections	1/2.5	1/3

The foundations of the work should also be stable from the mechanical point of view. One should be concerned not only with the fill material of the dyke, but also with the ensemble of the fill and the foundation. When the foundations are of bad quality, for example clayey, the slope of embankments is decreased by broadening the base.

I-7.2.3 *Protection by a system of internal drainage*

The objective of such a system is to reduce or annul the pore pressures in the embankment and prevent piping due to oozing of seepage water at different points on the downstream face of the dyke, which would be unfavourable to stability.

DRAINING THROUGH FLOOR CARPET

This is made with well-drained media (coarse sand) placed in the downstream portion of layers close to the foundation, covering 50 to 70% of the total width of the dyke at its base.

The efficiency of the draining floor carpet depends mainly on the hydraulic anisotropy of the body of the dyke, k_h/k_v (ratio of the coefficients of horizontal and vertical permeability); it is maximum when the body of the dyke is isotropic ($k_h/k_v = 1$).

CHANNELLED DRAIN, VERTICAL OR INCLINED UPSTREAMWARDS

This drain is attached to a horizontal draining floor carpet. It intercepts the entire flow of infiltration irrespective of the degree of anisotropy of the body of the dyke. The pore pressures downstream of the drain become null, thus eliminating piping, quicksand phenomena and erosion.

I-7.2.4 *Protection by impermeabilisation*

This is conceived as a watertight barrier in a medium which is too permeable or has too strong hydraulic non-homogeneity (dyke and/or foundation strata). This barrier may be placed on the upstream face of the embankment and eventually prolonged vertically across the foundation or it may be placed at the centre of the dyke, implanted in the axis of the body. The techniques of construction are varied:
— rigid or semi-rigid barrier: sheet of concrete or bitumen-concrete or soil treated to make it impervious;
— flexible barrier: plastic sheet, curtain of bentonite-cement, geomembrane or clay puddle.

These devices of cutting off seepage may also be used, when relevant, for lateral connection of the dyke with the terrain (e.g., with the slopes of a cliff).

I-7.2.5 *Complementary protection*

FILTER TOE

The drainage system is generally completed by a filter toe made of coarse material (gravel, blocks, rocks) and embedded up to about 1 metre in the foundation; water drains out of the filter toe near the outlet.

DRAINAGE DITCHES

There is a possibility of the foundation soil immediately downstream of the dyke rising and getting washed away due to water seeping from the retention basin through a layer whose permeability is larger than that required for reaching the surface and hence oozing out under pressure. This can be prevented by providing ditches or trenches to alleviate the extra low pressure at the base of the surface layer. The ditches or trenches are located at a significant depth in the foundation along the downstream face.

I-7.2.6 Filtering of drainage materials

The drainage system consists of materials of high permeability ($k \ddagger 10^{-4}$ m/s). This engenders the risk that seeping water may entrain fine particles of the surrounding soil. This phenomenon can be controlled by using proper material which will act like an inverted filter.

For these granular filters in direct contact with the soil, the rules most commonly applied are those proposed by TERZAGHI [54]. They often imply use of granular material prepared in specified factories. Development of geotextiles for various uses in civil and earth-moving engineering, facilitates the execution of works, in particular those related to drainage and construction of filters. Protection by filtering around a drain made of granular material can be obtained by laying a pointed or thermally welded plain geotextile at the interface between the soil to be filtered and the draining fill, except that for this purpose expertise in granulometry is needed. Also, the choice of the textile has to accord with the criteria stated in *Emploi des Géotextiles dans les systèmes de drainage et de filtration,** which contains recommendations made by the French Committee of Geotextiles and Geomembranes.

I-7.3   Examining Stability of the Dyke

I-7.3.1 Introduction

This study involves verification of stability of dykes or embankments at the most critical stage of their operation. It is quite simple and easy in the case of dry basins, or infiltration, or when the dykes and the underlying terrain are separated by a watertight geomembrane, or when the height of the dyke is very small (1.5 to 3 m) and it is located on a favourable site.

On the other hand, 'permeable' dykes of detention ponds, of relatively large height situated on an adverse site, require a profound study to justify the safety measures taken. Examination of stability should be carried out in short and long terms according to the methods prescribed by the mechanics and hydraulics of soils.

I-7.3.2 Various tests for stability

(a) EXAMINATION OF STABILITY OF THE FILL ON ITS SUPPORT
This study generally requires a computer or other computation facilities. The safety factor should never be less than 1.5. Should the case arise,

---

*Use of Geotextiles in the Drainage and Filtration Systems

compaction of the foundation soil shall be carried out according to the soil-consolidation procedures prescribed in the construction of embankments on compressible soils.

(b) EXAMINATION OF THE DOWNSTREAM FACE OF THE EMBANKMENT IN SERVICE
This study is executed under usual constraints and necessitates, as shown in Fig. I.15:

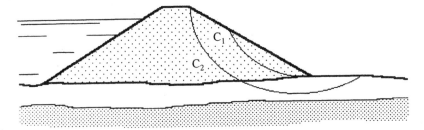

Fig. I.15 Stability by potential circles of slippage: C_1 is the circle of potential rupture of the downstream embankment, C_2 the circle of potential rupture causing cracks towards the downstream, affecting the foundation soil (very dangerous rupture resulting in total ruination of the work).

— Measurement of pore pressures in the dyke and its foundation in the case of steady flow at the normal water level or at the highest water level. Such measurement is necessary if the highest level is sustained for a long duration of time, of the order of fifteen days or more (common occurrence in detention ponds). Further, this becomes very important when the permeability of the dyke material and of the foundation is significant.
— Computation of stability by circular (and/or non-circular) rupture on the basis of the relevant coefficient of soil shear and the geometry of layers. The safety factor should be larger than 1.5.
The hydraulic model will also give the value of the discharge.

(c) EXAMINATION OF THE UPSTREAM SIDE OF THE EMBANKMENT UNDER
    CONDITION OF FAST EVACUATION OF WATER
The procedure is the same as in the preceding case and takes into account the water body being filled up to the highest level, and a distribution of pore pressures identical to that obtained in the case when the embankment is isotropic and flow over it horizontal, and the characteristics c' and φ' (cohesion and angle of friction) of the drained soil are uniform.

The simplicity of the hydraulic network consisting of stream lines and equipotential lines within the dyke facilitates computations. In this case the safety factor may be as low as 1.2.

CONSTRUCTION— INSTALLATION

II-1   CONSTRUCTION OF OPEN BASINS

This chapter focuses on dry basins and detention ponds, bounded or not bounded by a dyke. If bounded, construction procedures for detention ponds may differ from those for dry basins. Particular attention is paid to dyke construction. This point is project-specific. On the other hand, **guidelines with respect to soil, earthwork and watertightness are applicable to all types of basins including underground tanks.**

II-1.1   Complementary Studies

The first step is a thorough reconnaissance. One may recall the need for a very careful reconnaissance of the strata and subsoil under the hold of the proposed dyke and its environs. The procedural points were presented in Section I-2.4.2.

Frankly speaking, there is no typical reconnaissance plan; each project has its own characteristics originating from design and the site on which it is to be executed.

Furthermore, the scale and means of reconnaissance should be commensurate with those of the project. The strategy adopted in the case of a dyke less than 5 metres in height will differ from that adopted for a dyke around 15 metres in height, especially if in the first case the foundation soil and local terrain present no well-known problems regarding stability (homogeneous layers without water table being intercepted)

II-1.1.1   Exploratory drilling under dyke hold

As stated, the plan of subsoil investigations may differ significantly from one work to another; nevertheless, the plan and nature of test drillings should be, insofar as possible, based on the following principles:
— Exploratory drilling (core, percussive, destructive) should be localised along the axis of the dyke in the direction of a profile perpendicular to this axis at the point where the dyke is highest, with spacing between drill holes less than 50 m (see Fig. II.1).

— Exploratory test points (penetrometer, stress meter, strain meter) should be distributed along the axis and foot of the dyke.

Zones a priori more critical are those around the greatest height of the dyke and its extremities (risk of bypass), the areas of inclusion of concrete structures in the dyke. All these zones must be particularly investigated.

Mechanical drilling for general reconnaissance should be maximised by using piezometric tubes or drills or diagraphic tubes (nuclear diagraphy). As for the last, it should not be forgotten that the existence of local but pronounced variation in thickness and nature of the layers (lenticular formations for example) may be prejudicial to the work if this aspect has not been attended to prior to commencement of the project. Hydraulic tests and study of the parameters of drilling also provide useful information.

As a general rule, exploratory drilling and other on-site tests should attain a depth at least equal to the maximum height of the dyke. However, it is often necessary to reconnoitre the foundation soil to a greater depth for more precise knowledge of either the geology of the site, if this appears complex, or such mechanical characteristics as compressibility or swelling characteristics of deeper layers if these are likely to play a determinant role in the behaviour and functioning of the dyke. Such indeed has been evidenced in the case of compressible soils, fractured rocks and sandy or gravelly alluvia. A study of surface geophysics has proven an effective tool in planning judicious mechanical drilling and deciding depth of drill holes.

II-1.1.2 Survey of site and borrow areas for fill material

The essential objective of this survey is examination of the watertightness of the basin and suitability of the on-site soils for construction of the dyke.

Seepage flows due to infiltration need to be measured only if this parameter is an important enough element of the project to influence its performance, either through causing a hydraulic imbalance between inflow and outflow of the basin or engendering a risk of pollution of the groundwater. In the case of detention ponds versus dry basins, it may sometimes be useful to limit the percolation losses across the basin bed. Lefranc tests of permeability, conducted below the bed level of the basin through drill holes, help in determining percolation rates. Drilling is carried out along a fairly coarse grid of between 100 and 300 metres depending on the homogeneity of the site and surficial area of the retention basin.

For obvious reasons of economy the borrow pits from which material to be used in construction of the dyke is taken, should be located inside the perimeter of the project. Soil from these pits is studied through random samples most often taken with an auger from loose soil. A grid of the order 50 m × 50 m is acceptable; the depth of drilling is decided according to the thickness of the embankment and the position of the water table which, in principle, constitutes the limit. In rocky or semi-rocky terrains preliminary investigations are carried out by means of a percussion or hammer drill.

A geological section of the Maurepas (France) dyke is depicted in Fig. II.1, together with the loci of reconnaissance drill holes.

II-1.1.3 Laboratory tests

The samples of soil collected at the time of survey are analysed in the laboratory for identification and behaviour.

The objectives are the same as those of the survey: determination of the possible use of the earth as fill material, stability of works and permeability of the underlying soils.

TESTS FOR IDENTIFICATION
The objective is classification of the soil and hence these tests include:
— granulometry,
— liquid blue value, Atterberg scale, which determines the degree of cleanliness of granulated soils and the plasticity of fine soils (or of the fine portion of granulated soil),
— fraction of organic matter (if relevant).

SPECIFIC TESTS ON FILL MATERIALS: CHARACTERISTICS OF STATE OF MATERIALS
— water content
— Proctor test,
— punching immediately or after saturation with water in a CBR press. These tests provide indices (may be termed load bearing indices) which indicate the quantum of moisture content from compaction corresponding to the maximum load-bearing capacity of the soil. Variation of these indices from those obtained from the Proctor test and the values of moisture content at the site help in deducing the state of consolidation of the soil at the borrow pit.

SPECIFIC TESTS FOR CALCULATING STABILITY OF WORKS
— specific gravity, both wet and dry;
— simple compression;
— traction test (Brazilian test). This test is conducted especially on rock samples or test cylinders of soil treated with cement;
— shear test: this helps in determining the cohesion and angle of internal friction of a soil;
— consolidation test: oedometric curves enable determination of parameters of soil compaction required for computation of consolidation under load and related time of consolidation;
— state of rock fracture: measured by the RQD (Rock Quality Design) index;
— rock hardness: determined by tests of strength under simple compression, tensile strength and normalised tests of wear such as the Los Angeles and Deval (or micro-Deval) tests.

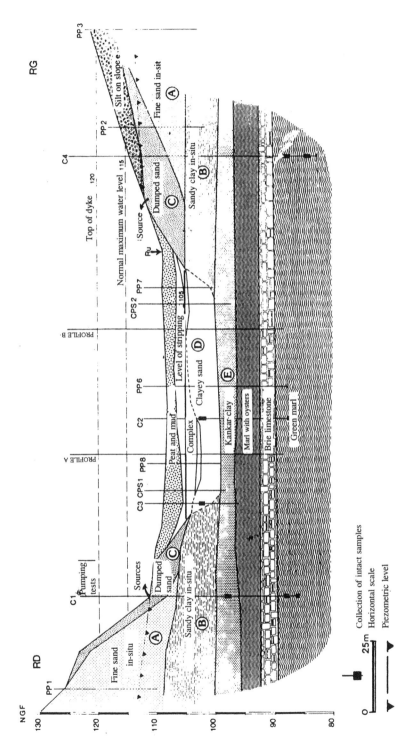

Fig. II.1 Geological section of the Maurepas dyke and plan of reconnaissance drill holes.

TESTS FOR MEASURING THE COEFFICIENT OF PERMEABILITY *k*

Like the Lefranc test for permeability of strata on site, the laboratory tests of permeability are carried out under a varying or constant head of water.

For all the above tests, it is beneficial to consult the manuals of geotechnical tests and publications of the French Ministry of Agriculture [54].

II-1.2   Construction of Earth Embankments

II-1.2.1   Initial preparation

Early actions before starting earthwork include cutting down trees, removal of stumps, demolition of any structures, displacement of any pipe or drainage networks, installation of a yard (including laboratory) and stores for needed materials.

Initial preparations for earthwork, embankments and borrow pits include:
— stripping and separately storing top soil;
— eventual diversion of watercourses on site and evacuation of stagnant water;
— fastening embankment or dyke on its base rock or laterally with the abutments through appropriate key blocks;
— elimination of layers or portions of layers in the foundation of the dyke found to be of unsatisfactory quality (excavation and replacement with material of appropriate quality);
— filling voids in the terrain by grouting (soluble rocks, crevices);
— levelling and evening out areas of embankments and borrow pits.

II-1.2.2   Earthmoving

Instructions given in the technical guide prepared by LCPC*-SETRA in France, relating to the 'construction of embankments and formative layers', consisting of two volumes: volume I — General Principles, and volume II — Technical Annexure (GTR, September 1992), can serve as a model for compilation of useful guidelines [43].

Although these publications are oriented towards construction of road embankments, an earthen dyke is very similar, even identical to road embankments and the methods and means of construction are the same in both cases, except for some adaptations in the case of dykes due to the hydraulic aspects of the work. The nature of the soil will often affect the design and structure of the dyke (seepage protection, upstream pitching, filtering massif, draining ditches etc.).

The aforesaid documents should be consulted for all questions concerning classification of soils and rocks and conditions of utilisation of soil and compaction of embankments. Here, only questions pertaining to the extraction of materials and construction of homogeneous dykes are taken up.

---

*Laboratoire Central des Ponts et Chaussées, Ministry of Public Works

II-1.2.3   *Extraction of materials*

CASE OF LOOSE MATERIALS (SAND, SOIL)

Depending on the category of soil and its state of wetness and consolidation, the *Technical Instructor* recommends two methods of extraction:

— Extraction by layers: this method favours surficial moisture evaporation. It is recommended in the case of wet soils, provided circulation on the soils is possible, and also in cases when separation and grading of materials are desired at this stage. The most suitable machinery includes scrapers, bulldozers and dumpers.

— Frontal extractions: this method is preferred during the rainy season as it limits the surface exposed to inclement weather and consequently prevents the adverse development of hydric soil. Shovels and dumpers are suited for this job.

CASE OF COMPACT ROCKY MATERIALS

Smashing: The 'rippability' of rocks may be defined as their aptitude for breaking under the action of the teeth of excavation machinery before being loaded into the usual earthmoving carriers. Successful ripping depends on several factors:

— the yard: ensuring adequate output and obtaining desired sizes.

— the rock: its nature and discontinuities may be of stratigraphic (superposition of layers of different hardness) or tectonic (cracking) origin;

— the material and its working conditions;

Use of explosives: When a rock is too compact to be ripped, explosives are used which instantly release large amount of energy. Control of this energy is an art and needs specialists capable of:

— properly distributing the energy generated in the rock by the explosion;

— determining the correct quantity of explosives;

— controlling the region affected by the explosion by adjusting the drilling size and intensity of explosion by means of retarders.

Use of explosives has, for several reasons, an impact on earth moving projects:

— at the level of geotechnic studies: it is useful to gather maximum information about the stability of the rock massif and the likely difficulties regarding the use and efficiency of explosives;

— at the level of geometry: extraction of large depths is executed in benches (about 15 metres high) to maintain good control over drilling operations; it is evidently prudent to take this aspect into account for positioning eventual berms. Also, as correct drilling is difficult when the slope is smaller than 2/1, the embankments of small slopes are obtained by a series of vertical operations, resulting in a somewhat uneven shape;

— at the level of reuse of the excavated material in a fill: to be reuseable, the rock must be so fragmented that the size distribution is compatible with the planned use and the equipment available for construction;

— lastly, at the level of the environment: the operation of explosion is accompanied by tremors, noise, flying stone pieces and dust. Thus in using explosives on the work site due care must be taken by the enterprise as well as the chief of works.

The technical Instructor, LCPC-SETRA [45], pertaining to 'removal of rocks by using explosives for building roadways' should be consulted by personnel involved in this type of work.

Earthwork below the water table: There are certain constraints associated with earthwork below the water table. The technique of 'digging' may be chosen in this case from various methods widely discussed in the technical literature. The reader may consult specialised publications: the choice of one or several methods takes into account the geologic and hydrogeologic context specific to the problems posed by the project. This technique of digging is sparingly used for earth moving or building below the water table. Also, the techniques need accurate data for correctly dimensioning the devices.

The first thing to look for in such work is the existence or creation of an outlet for evacuation of the pumped water. The important methods used are as follows:

— ditches and trenches for draining;
— excavation by pumping out from a caisson or coffer-dam and access pool;
— lowering the water table through drainage points, pumping from a sump well.

II-1.2.4 *Utilisation of excavated earth in embankment or dyke*

In most cases, the excavated earth will possess satisfactory mechanical and hydraulic characteristics for the construction of a dyke. In the contrary case, it may be suitable only for limited specific usage: lining of banks and bottom of the basin etc. The recommendations of the *technical Instructor* LCPC-SETRA [43] are useful here.

II-1.2.5 *Construction measures for homogeneous earthen dykes: Compaction and treatment of the on-site soils*

The work may be implemented and controlled in the same manner as a road embankment except with regard to hydraulic aspects. It has to satisfy in certain cases some specific constraints:

— Case of homogeneous dykes and dams with total hydraulic isolation. The work has no hydraulic function and, from the technical point of view, is entirely comparable to a road embankment. The choice of material is also the same.
— Case of dykes or dams with a hydraulic function. Infiltration across the body of the work and the bed is permissible. The choice of material

is no longer dependent on the conditions of compaction only; it has also to take into account, for its stability, the initial hydraulic conditions and the modifications caused by the work itself on the site (permeability of embankment, draining and waterproofing devices). The design of the work and choice of material are thus closely related in this instance and utilisation of soil available in the hold of the project remains a priority consideration.

The method of control, termed Q/S and described in the GTR, enables correct dimensioning of the compaction equipment as a function of the quantity of materials added daily and fixes the thickness of individual layers to ensure the quality of compaction desired. On the other hand, Q/S gives no information about physical parameters of the compacted soil. Such information is obtained by complementary measurement of the state of consolidation (ρ_d) by gammadensimetry in-situ and in the laboratory by Proctor tests which determine the rate of consolidation. These parameters of state of soil together with those of identification, constitute an essential objective for field studies and are directly related to the hypotheses formulated during computation. The generally required rates of consolidation should be higher than 95% of the OPN (Optimum Proctor Normal)

The machinery most appropriate for compaction of plastic soils are sheepsfoot rollers or static rollers with vibrating spikes when their scope of use correctly corresponds to the materials to be compacted (see tables of use compiled by GTR). This is critical since homogeneity and mutual cohesion between layers lead to phenomena of hydraulic anisotropy ($k_h/k_v \neq 1$), always harmful to stability in the case of a hydraulically non-isolated dyke.

The technique of in-situ treatment of soil by hydraulic binders (lime/cement) might possibly be considered in the construction of earthen dykes when problems of compacting (very wet soil), permeability (reducing the coefficient of permeability) or stability (augmentation of mechanical strength) arise. The treatment should not introduce hydraulic heterogeneities in the work. The functional part of the hydraulic zone should be completely treated even if the basic objectives which prompted implementation of the technique are modified during its execution.

Very dry and treated soils increase the rigidity of the body of the dyke and consequently its sensitivity to unequal settlements, which can give rise to a network of cracks, very damaging for the works. This happens especially in the case of highly compressible soils in the foundation or in dykes of very large height. This aspect must be taken into consideration while formatting the project or during implementation after evaluating the eventual effects on its good performance. Many authors, keeping the aforesaid in mind, have recommended that the moisture content in fill material used for dykes be kept somewhat higher than the optimal Proctor value as this would improve the flexibility of the structure. At the same time, the moisture content must remain within the limits necessary to satisfy the specifications of consolidation and preclude development of interstitial pore pressures in the structure.

II-1.3   Watertightness

For various reasons relating to the nature of soil, the stability of works constructed and/or the protection of the natural environment, it may become necessary to have recourse to one of the methods of creating watertightness described below. This may be of concern for dry basins (case of an aquifer close by) as well as detention ponds (case of deficient balance of water and/or protection of the water table in the vicinity). Hollow basins like those bounded by a dyke may also fall in this category.

II-1.3.1   *Treatment of in-situ soils: watertightness by mechanical and chemical measures*

COMPACTION
This is the simplest way to reduce the permeability of soil. This action is necessary for all construction work in soil but the nature of the soil strongly determines its effectiveness; compaction has greater effect on the permeability of gravel than that of fine soil, all other factors being equal (intrinsically, a fine soil has lower permeability than a granular one).

ADDITION OF FINE MATERIAL TO THE NATURAL SOIL, CLAY PUDDLING
One of the most common techniques for reducing the permeability of a soil consists of adding clay to it (for example montmorillonite or bentonite). Thus the value of the permeability of muddy sand is reduced to one-tenth or one-hundredth by adding 10% bentonite.

A primary method consists of drying, cleaning and levelling the zone to be treated at the bottom of the basin, then covering the holes and cracks with a mixture of 1 part bentonite to 5 parts earth. Subsequently, this layer is plastered to a thickness of 15 to 20 cm and, after drying, the upper surface smoothened. Next, bentonite is spread over the surface at the rate of 4 to 10 kg/m^2 depending on the permeability desired. The mixture is made to penetrate to a thickness of 10 to 15 cm by kneading and the layer compacted with a roller.

Another method consists of first levelling and compacting the bottom, then spreading a layer of pure bentonite, which is subsequently covered with a 10 to 15 cm thick protective layer of sand or gravel; the quantity of bentonite used is about the same as in the preceding method.

Nowadays a prefabricated product consisting of a thin layer of bentonite with protective coatings of geotextiles on both sides is available.

Every clay puddle must be quickly protected from desiccation, as this inevitably leads to cracks, thereby annulling the desired effect. Clay shrinkage is particularly critical in the case of masonry structures.

USE OF CEMENT AND OTHER BINDERS

Widely used in sole layers of roads, this technique can be employed in the case of retention basins. Apart from its effects on permeability, this treatment significantly improves the suitability of the fill material for construction and maintenance of basins.

Treatment of soil by adding bentonite or its on-site treatment with cement or quick lime in reasonable proportions leads to lowering of the coefficient of permeability to one-tenth or one-hundredth for granulated or muddy soils. A study for determining the appropriate quantities of binders is recommended.

In this context, mention may be made of rendering a soil or rock impermeable by injecting grouts through the bottom of a basin and of creating grout-curtains across the dykes through injection. For this operation to be beneficial, the initial permeability of the rock (due to cracks) and the soil (sand and alluvial gravel, rocky blocks) must be high. A soil with a permeability of $k \leq 10^{-5}$ m/s is considered non-injectable; further, the closer the permeability to this value, the less injectable the soil.

OTHER TECHNIQUES

There are, of course, other techniques for producing watertight surfaces, namely spreading a layer of bitumen-concrete or hydraulic cement-concrete. Such procedures can be used for the bottom of basins as well as for dykes. They should be implemented on drainage passages to reduce pore pressures and on supports under large loads.

These rather expensive coatings are preferred in cases where the bottom of a basin is subjected to significant mechanical loads, in particular maintenance machinery.

II-1.3.2   Watertightness by geomembranes

CHOICE OF APPROPRIATE MEMBRANE

The choice of membrane (nature and thickness) depends on the characteristics and constraints of the site, such as the nature of the foundations, mechanical stresses, exposure to light, nature of collected water etc. Other factors affecting choice are the conditions under which the membrane will be used and the cost.

Table II.1 lists the materials most widely used in the manufacturing of geomembranes during the last decade (1980-1990); the list is not exhaustive.

INSTALLATION OF GEOMEMBRANES

In applying a geomembrane for watertightness, a stable support of high quality smoothness and no direct contact with elements likely to damage it must be ensured (Fig. II.2).

Table II.1   Geomembranes used for impermeabilisation of earthen structures

Chemical composition	Commercial name	Thickness (mm)	Tensile strength (MPa)	Extensibility (%)
PVC (polyvinyl chloride)	Eurofolor (reinforced polyester) Sarnafil Rhenofol	1.0	12	150 (PVC)
	Intertem WBF	0.6	15	200
	Drakatechnofol 0411	0.5–1.6		
	Sika-Norm			
	Wolfin			
	Trokal WB (Dynamit Nobel)	3.0	20	250
E.P.D.M. (mixture of ethylene, polypropylene and diene)	Hertalan Hermesit Sika-Teraplan Miner-Eutyl-Membran	1.0–2.0		
I.I.R.-isoprene isobutylene rubber butyl diene	Rhepanol Butyl (Esso) RMB (Esso)	0.5–2.0	10	400
Bitumen tissues	Hypofors (AKSO-ENKA nylon tissue)	3.0–5.0	9-23	25
	U.P.M. (polyester tissue)	1.5–5.0		
	Lucobit 1210 (glass tissue)			
	Coletanche NTP3 (felt)	5.0		
Synthetic elastomers	Hypalon Hy-bar Nitril-Rubber	0.5–2.0		
Polyolefine	3110 Du Pont	0.5–1.5		
E.C.B. (copolymer ethylene bitumen)	Carbofol Tixproof Aquagard Bitulen	1.0–2.0		
	Lucobit (Niederberg Chem.)	2.0	35	700
P.E. (poly-ethylene)	Schlegelplatte Saraloy (CPE-PE and PVC)	1.0–2.5	8	500
	Trespalen (H.P.D.E.) (HOECHST-GUNDLE)	1.0	20 to 25	600

Slopes of surfaces treated for watertightness: The characteristic mechanical stability of the protective device should satisfy the same conditions as envisaged for the body of the dyke, taking into consideration such adverse situations as local failure of the watertightening or saturation of the soil in the vicinity of the membrane. The choice of slope should take into account the difficulties of installation of the watertightening system. Slopes of 1/1.5 or higher are permissible only if they satisfy the conditions of stability of the

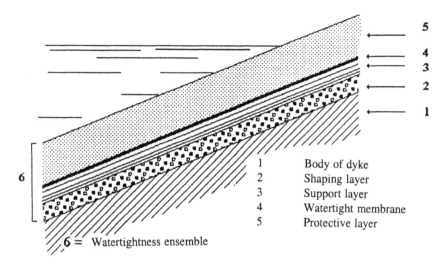

1	Body of dyke
2	Shaping layer
3	Support layer
4	Watertight membrane
5	Protective layer

6 = Watertightness ensemble

Fig. II.2 Watertightening of surfaces (case of a dyke).

work. Smaller slopes, say 1/2.5, largely simplify the preparatory works and application of protective coatings.

Support of watertightening layers: Such support is usually also made to serve as a filter for the bottom of the basin and/or the body of the dyke. The support layers may be composed:

— either of gravelly sand per se or of washed granules less than 20 mm in diameter without fine particles; the thickness of this layer should be at least 15 cm; stabilisation with application of a surficial emulsion is recommended for slopes steeper than 1/2.5;

— or of fabric-backed bitumen or open and draining hydraulic cement-concrete in 10 cm thick layer;

— or a composite geotextile with hollow or honeycomb core capable of supporting the installed watertight membrane.

The coefficient of permeability of the material should be larger than 10^{-4} m/s.

For simple cases (bottom, embankment or dykes of small height) the supporting layer may be reduced to a simple layer of transition or to a layer of a non-putrescible geotextile.

In steep parts, from the hydraulic point of view, this layer may be considered a draining chimney linked to a hollow drain at the foot of the embankment or dyke. This arrangement avoids much trouble during quick evacuation by eliminating pore pressures harmful to maintaining the membrane on its support (Fig. II.3).

Installation: It is useful to chalk out a definite plan of installation of watertightening in terms of the geometry of the basin and the width of bands, with each element located and numbered: installation of a membrane is a job for specialists.

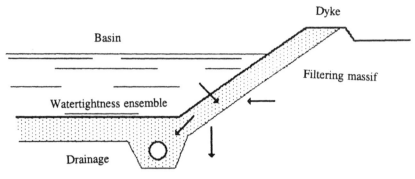

Fig. II.3   Effects of drainage of water behind a geomembrane.

Any risk of accumulation of gas and water under the membrane should be eliminated by arranging exit points for gas at summit points and those for water at trough points. Precautions have to be taken at the level of design, especially in the design of appropriate slopes for the bottom of the basin to facilitate de-aeration near the edges, and in the limitation of horizontal joints on the slopes of the dyke. Recourse to a system of draining network (PVC drains or composite geotextile) is sometimes necessary to expedite gas evacuation for example in peaty or fractured rock zones. One has also to guard against the phenomenon of piston effect due to fluctuations of the groundwater table if located close to the bottom of the basin; such fluctuations during the water table rise may force the air in the soil towards the membrane, thereby generating forces equivalent to or much larger than those of gas emission. There is risk of the same phenomenon occurring under the hydraulic effect of upthrust in both cases, vents and exhausts are necessary. From these anomalies emerged the device humorously called the 'hippopotamus' for watertightening surfaces of water bodies.

Pressures generated by air cause significant difficulties in the installation of membranes and excess residual stress observed in the upper parts or vicinity of anchorage. For large works, a reinforced zone (extra thick, reinforced membrane) should be planned.

Anchorage at top and bottom is generally provided in a trench or by a specific work ensuring continuity of watertightness, especially at the bottom. Continuity of the membrane, while spreading it under the basin, does not obviate the need for treating portions near the bottom as specified above. Provisional anchorage is sometimes useful for good stretching of the membrane by successive recovery at the level of folds and precludes tension in some materials with a large coefficient of dilation.

Applying a watertight membrane on a work constructed of concrete requires the interposition of a thick coating of material with good aptitude for deformation under pressure, together with chemical compatibility with the membrane. In the case of coatings of the elastomer-plastomer type, metallic plates complete the arrangement for installation. These points of linkage should not become the seat of excessive stresses likely to cause failure or weakening of the membrane by stretching, following, for example, excessive compression of the structure on which the membrane is anchored.

Jointing of membranes and control of watertightness: During jointing two membrane sections, a few centimetres overlap in the case of synthetic materials and about twenty centimetres overlap for bituminous materials, is provided.

Mode of jointing: Current modes of jointing include:
— thermal welding with or without addition of material (single or double welding),
— welding using solvents
— joint by means of an adhesive, viscous liquid applied under hot or cold conditions,
— joint by means of an adhesive film,
— vulcanisation.

The joints are welded automatically or manually. The automatic welding over a large length is generally of better quality because the parameters of pressure and velocity are better controlled. Before welding, the edges should be prepared by filing or cleaning.

Control of the quality of joints has to be exercised both for continuity of weldings (watertightness) and their mechanical strength. Control of continuity may be tested by non-destructive methods:
— visual examination (particularly effective in the case of translucent material),
— passing a pointed tool (along the external edge of the joint),
— blowing compressed air (minimum pressure 500 kPa),
— determination of thickness by ultrasonic analysis,
— vacuum chamber,
— subjecting the central line of a double welding to hydraulic or pneumatic pressure,
— electrical method: dielectric snap by passing current in case of discontinuity along a conductor wire glued along the interior edge of the joint.

Stength of joints is tested by destructive tests:
— Test by traction: this consists of subjecting a test piece, slit perpendicular to the joint, to a unidirectional traction perpendicular to the plane of the weld. Test pieces in the form of bands are 15 to 30 mm wide; the rate of stretching is between 50 and 200 mm/min. Failure often occurs not at the joint, but on the geomembrane itself in a zone of weakness. The reference values of yield stress are

$$4 < R < 6 \text{ N/mm} \qquad \text{good strength}$$
$$R > 6 \text{ N/mm} \qquad \text{very good strength}$$

— Contrary to the preceding case, test of shear strength is conducted in the plane of the weld. The strength may be termed `good' when failure occurs at the level of the joint. An index of welding compatibility is also determined; it is defined as the ratio of the strength of a joint to that in the standard part of the membrane.

Protective layer: This is the last layer placed above the watertightening layer or membrane to protect it against mechanical, physicochemical, or solar radiation attacks. The interposition of a geotextile between the membrane and the layer itself is often necessary for installation and better bonding of the layer (increasing the coefficient of friction).

As in the support layer, here, too, the materials used in the sloping surfaces should have a large coefficient of permeability ($k > 10^{-4}$ m/s). This can be achieved by:

— rocks arranged manually,
— porous bitumen-concrete,
— permeable cement-concrete (example: gravel 1350 kg/m^3, sand 200 kg/m^3, cement CLK 330 kg/m^3, water 180 kg/m^3 + plasticisers),
— three-dimensional geotextile in which stones fill the alveoli,
— simple layer of gravel or stones (slope < 1/3).

The plane portions of the bottom of a detention pond may be left uncovered, without protective layer: the water itself plays the protective role. Maintenance operations may, however, cause damage; a protective coating of mud is therefore recommended.

The various techniques used for attaining watertightness at the base and sides of a basin are summarised in Table II.2.

Table II.2 Devices for watertightness

Techniques	Recommendations
Compaction	Indispensable for construction by earthfill laid in layers. More effective for granular materials than for fine soils, all other things being equal.
Clay puddle and soil treatment with hydraulic binders (active lime or cement)	Useful if materials of good quality are available on-site or nearby. Attention should be paid, however, to the phenomenon of shrinkage, especially for the lower layers. In-situ treatment of soil with hydraulic binders enables utilisation of humid materials, precludes earth movement and allows resistance to loads.
Bitumen-concrete and hydraulic cement-concrete	Recommended if the bottom of the basin has to bear large mechanical loads. Due care must be exercised for this kind of coating, especially in the case of joints, to ensure desired watertightness.
Geomembranes	Watertightness almost total. Implementation delicate with need for follow-up control (especially of welds). Attention to be paid to pressure variations on each face of membranes.

II-2 OPEN BASINS: ANCILLARY WORKS

II-2.1 Inlet Works

Direct debouching of conduits, especially those of large diameter, into the basin is not recommended; the aesthetic effect is not pleasing. Conduits are best concealed in the body of the embankment, thereby giving them a 'bayonet' profile and facilitating debouching below the water level. If a remediation unit is to be located on a conduit, it is placed upstream of the point of outfall. Some trees, in particular weeping willow, help in ready concealment of conduits, making them almost invisible (Fig. II.4).

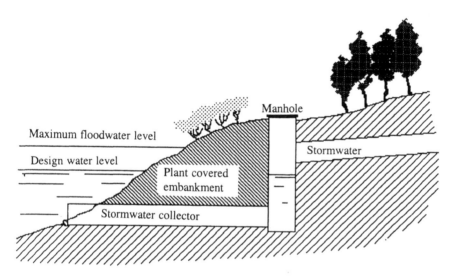

Fig. II.4 Simplest case of inlet of stormwater without a unit for remediation.

II-2.2 Pretreatment Works

II-2.2.1 Introduction

Flow of stormwater in urban regions entrains significant pollution; this was discussed in Chapter 1: Design and Planning. It may be recalled that with respect to suspended matter, a large part of this pollution can be easily removed.

Installation of pretreatment units is gradually becoming a standard practice, not only to protect retention basins, but to alleviate concern for the receiving waters and catchment areas coming under accidental or chronic pollution. These units, single or arranged in series, are usually installed upstream of the retention basin and only rarely downstream. As for their dimensioning, it is useful to consult such reference texts as *Memento Technique de l'Eau Degrément* [53].

II-2.2.2 Screens

The objective of screens is to retain the macrowastes (plastic bottles, leaves, paper) likely to disturb the functioning of downstream works (pumps, evacuation devices etc.) and to render the stormwater 'aesthetic' for the natural environment (see RANCHET and RUPERD [64])). A screen made of equidistant bars is placed across the flow, thus holding back all materials larger in size than the spacing between bars. Current practices in screen installation are given below:
- spacing between bars (8 to 80 mm) depending on use to which the drained zone is to be put and frequency of maintenance,
- shape of screens (curved, straight),
- position of screens, vertical, inclined (60-80°),
- number of screens in series (limited to 2),
- mode of cleaning screens:
  - manual scraping: implying frequent cleaning,
  - mechanical scraping: in this case the weight of the screen, which increases with waste deposition, automatically activates scraping by a mechanical rake. This installation requires an electrical power source and regular maintenance.
- barred section (circular, rectangular, trapezoidal); according to some researchers, a trapezoidal section with a large base facing the flow is the most effective.

In addition to the parameters of construction (spacing and shape of bars), screen efficiency depends on the approach velocity of water. A large velocity of approach increases the accumulation of wastes inbetween the bars and screen clogging ensues. It is thus advisable to provide a straight approach channel capable of distributing the velocity uniformly over the ensemble of screens. Further, a trap for stones located upstream of the screens will prevent damage to the bars from carried stones. Lastly, to obviate screen clogging, a bypass may be installed upstream of the screen or a section without bars provided in the upper part of the screen to facilitate water flow.

II-2.2.3 Grit removal

Grit removal seeks to trap suspended particles larger than 200–250 μm in size, mostly gravel and sand, whose accumulation can reduce the storage capacity of retention basins in the long run. It also protects the works situated downstream, especially pumps.

Whatever the design of the degritting units, particles are trapped due:
- either to a reduction in horizontal velocity of flow by increasing the transverse section of the container, thereby causing settlement of grit by gravity (case of longitudinal grit chambers);
- or to introduction of a cyclonic chamber in the inlet channel in which the sand is deposited under the combined action of centrifugal forces and those of gravity (case of cyclonic degritting units).

Longitudinal grit chambers are more commonly used for treatment of stormwater. They consist of long basins with horizontal flow and are often equipped with a pit in which grit collects (Fig. II.5).

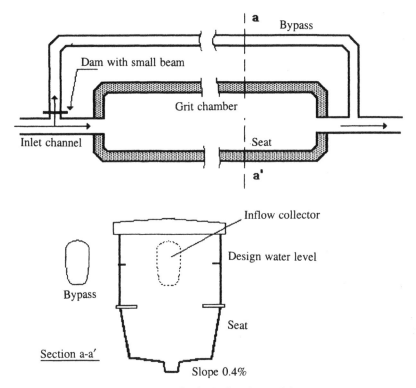

Fig. II.5   Longitudinal grit chamber with bypass.

Dimensions of grit chambers are determined by application of the laws of sedimentation. In practice, by enlarging the section of the inlet channel, the horizontal velocity of water is reduced to a value between 0.2 and 0.5 m/s (0.4 m/s in the case of urban stormwater). A lower velocity would provoke deposition of particles whose density is smaller than that of sand while a higher velocity might not ensure adequate deposition of grit (see [63]).

Degritters are generally designed for the peak flow corresponding to light rains. For peak flows of heavy precipitation, works of very large size would be required.

The most widely used longitudinal grit chambers have a rectangular section. Their performance is optimal only for the flow rate for which they are designed.

Various alternative devices help to maintain a constant (or only slightly varying) flow velocity in the grit chamber irrespective of the quantity of flow to be treated. Such devices improve the performance and also limit the deposit of very fine particles (organic matter which emits foul odours due to fermentation) during low flows. Among these devices, one may mention in particular:

Proportional flow weirs (Q = Kh): Installation of a weir satisfying a linear 'flow-to-depth' relation downstream of a grit chamber of rectangular cross section results in a discharge which varies linearly with water depth (h) in the grit chamber; the flow velocity thus remains constant.

Grit chambers of parabolic cross section: In these units, the part of the channel reserved for settling grit (useful length) is protracted by a narrowing device (venturi) which terminates in a vertical pit acting as a weir. The rate of discharge in this case is a power function of the water depth (h) upstream of the pit:

$$Q = 1.7 \times s \times h^{3/2} \qquad (Q \text{ in } m^3/s)$$

if the section of the channel is parabolic in shape. This expression is valid only for maximum flow rates of 500 to 600 l/s. In effect, the constraints of construction of the venturi (angle < 7° with the longitudinal wall) result in excessively large dimensions of the works for large flow rates (Fig. II.6).

The efficiency of grit chambers depends on:
— flow rate for which they have been designed,
— their position in the storm sewerage network,
— their maintenance.

Several comparative studies have been carried out on this subject in the last few years and the results are presented in Table II.3.

Table II.3   Average efficiency (as percentage) of grit chambers [71]

Type of device	SS	Mineral matter	BOD$_5$	CQD	Pb	Zn
Ordinary grit chamber[1]	8 to 22	18 to 60	—	—	—	—
Grit chamber with constant flow velocity[2]	62	(80–95)	35	43	58	42

[1]in combined sewerage network
[2]in stormwater network

The efficiency of grit chambers can be improved by decreasing the flow velocity flow upstream of the device by placing obstructions (vertical bars to break the flow) to generate head energy loss. Or (flat) plates may be placed which help trap solids without increasing turbulence which has an adverse effect on settling. The re-entrainment of particles is limited by by-passing the surplus flows or by using designs that maintain constant flow velocity.

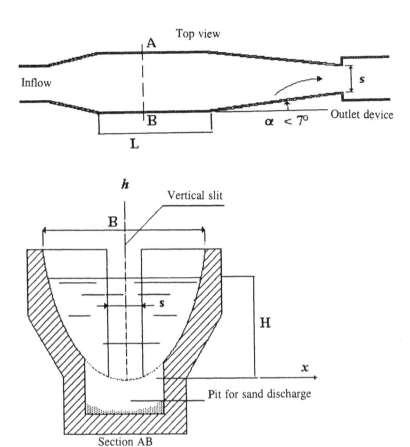

Fig. II.6   Sketch of a grit chamber with parabolic section.

Grit chambers may be designed as single units or double compartments with the possibility of isolation so that the compartments can act in turn. For stormwater systems a single unit suffices because maintenance can be carried out during the dry season.

II-2.2.4   Oil and grease removal units

The objective here is to eliminate the oil and grease which spread in a thin layer on the water surface and hamper its reoxygenation by slowing down diffusion of air. In addition, such removal units help in the recovery of other particles and materials of density less than that of water (floating objects, polystyrene etc.) which are sources of aesthetic nuisance and are not acceptable to the public.

Nevertheless it may be emphasised that except for accidental pollution or particular cases (parking lots, industrial zones, airports etc.), the quantum of oil and grease in stormwater is small: it rarely exceeds 50 mg/l and is often lower than even 10 mg/l (average over several stormwater samples). Also, most of the oil is adsorbed on the suspended solids (see Table I.4). Settlement for a few hours in a settling basin thus helps eliminate a large portion of oil and grease—69% on average (BACHOC and CHEBBO [1]).

In most cases the deoiling of water is effected by natural flotation (deoiling by gravity). For the flotation process to be efficient, the flow velocity must be limited in order to obtain a velocity of 'settling' (or rising) of 6 to 18 m/h, depending on the density of oil and grease (0.95 to 0.85).

Deoiler with Scumboard

The simplest and most widely used procedure consists of placing in the settling unit or upstream of the retention basins, a hanging partition or floating barrage which retains the oil and grease floating on the water (Fig. II.7). To remove the collected oil and grease from the device, several procedures are used:

— pumping,
— decanting the upper layer of liquid through a suction device before storage,
— installation of a special device for recovery of material floating on the surface (for example the Nenuphar system of ALSTHOM FLUIDES SAPAG).

One may also apply absorbents such as synthetic tissues but this procedure is rarely used today. A study carried out by L.R.O.P. on a deoiler with scumboard showed a retention of 59% of oil and grease (weighted average) during the storm period (RUPERD [73]).

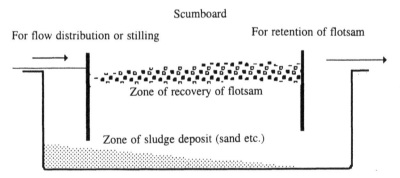

Fig. II.7   Sketch showing principle of a settling-cum-oil-and-grease-removal unit with scumboard [64].

DEOILER WITH FLOW BREAKERS

Concrete flow breakers are erected in the unit to retard the flow and create calm zones which enable oil and grease to rise to the surface in a rather short time (70–80 seconds for the maximum projected flow of 2.5 m^3/s). Flow-breakers significantly reduce the size of the work; the rising velocity taken for design may be of the order of 40 m/h. Some units designed by the firm BERTIN are in service in France, notably at Saint-Quentin-en-Yvelines (Table II.4).

Photo 5   Deoiler with flow-breakers at BOIS-ROBERT, France [72].

Table II.4   Efficiency of removal of pollutants by deoilers with flow breakers at BOIS-ROBERT, France ([RUPERD [72])

BOD$_5$	COD	SS	Oil and grease	Lead	Zinc
33%	32%	21%	53%	12%	10%

CLEANSER DEOILER (Fig. II.8)

Prefabricated units, marketed notably by the firms Saint Dizier, Separepur, Itera, Simop, Bonna etc., can treat large discharges (up to 500 l/s). The average removal is 47% for oil and grease, 14% for suspended matter and 24% for BOD$_5$ (RUPERD [70]).

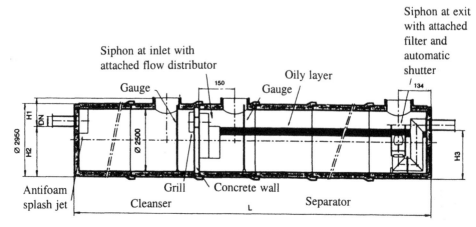

Fig. II.8   Sketch of cleanser-deoiler [64] (BONNA document).

II-2.2.5   *Settling tanks*

Settling tanks differ from grit chambers only in terms of size of particles removed. As a matter of fact, even settling tanks remove only non-soluble mineral and organic particles larger than 20 μm in diameter. Their cleansing performance is rather significant; it has been observed that settling for a few hours markedly reduces the quantity not only of SS, but also of the material attached to it (see Table I.8).

Settling tanks work on the same principles as grit chambers. Their dimensions are determined from the usual laws of sedimentation.

RECAPITULATION OF SETTLING
Some general concepts regarding settling which are essential for understanding the related phenomena are discussed below; they are developed in a theoretical and simplified manner. It is assumed that the devices are considered as ideal settling tanks, i.e., the velocity field is uniform over the entire wetted section and the flow is of the piston type.

A sketch of the working of a settling tank is given below.

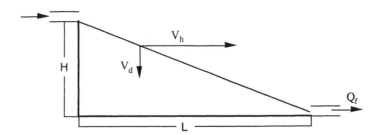

Here V_h is the horizontal velocity of particle movement; V_d the velocity of settling or the vertical velocity of the particle; Q_f the release rate through the settling tank.

In order for settling to take place, i.e., a particle to be intercepted by deposition on the bottom of the basin, the time of horizontal travel t_h must be equal to or greater than the time of settling t_d. L is the length of the device; H its effective depth (maximum depth to which it is filled); t_h the time taken by the particle to travel the distance L; t_d the time taken by the particle to reach the bottom of the basin (i.e., to travel the vertical distance H).

The velocity is assumed to be uniform over the vertical section of the device. Thus

$V_h = Q_f/S$ where the cross-sectional area of the basin S = B × H,

B being the width of the basin and S the vertical section of the work. Hence,

$$\frac{V_h}{L} < \frac{V_d}{H}, \quad \text{or} \quad \frac{Q_f}{B \times H \times L} < \frac{V_d}{H} \quad \text{so that} \quad V_d > \frac{Q_f}{S_h}$$

S_h being the surface area of the basin or $S_h = B \times L$; V_d is called the Hazen velocity.

The particle is therefore intercepted if the settling velocity V_d is higher than or equal to the quotient of the release rate and the horizontal surface area of the device.

There are two types of settling tanks:
— standard settling tanks similar to those existing in water-treatment works,
— lamellar settlers also called on-line settlers.

STANDARD SETTLING TANKS

These may be rectangular or circular in shape; in both cases the flow of water is horizontal.

Further, the horizontal velocity must be sufficiently low. To ensure adequate stability of settling over a large range of regimes, it is recommended that the ratio width/depth be equal to 1.5 and the useful depth be 2.5 to 3 m.

The ideal settling tank does not exist in practice, however. Some deviations of a practical settling tank from the ideal basin are listed below:
— the flow of water is not homogeneous (vortices, dead zones etc.) and hence the velocity is not uniform over the vertical sections;
— reverse flow called recirculation, often occurs in some regions; consequently, only part of the volume is useful;
— the ideal model does not take into account the fact that solid particles, once deposited, may re-enter the flow;
— the assumption that the velocity of practical settling to be more or less equal to the Hazen velocity of settling is a gross approximation; the

circulating flow and turbulence due to differences of temperature and density always lead to certain disturbing transport phenomena in the flow; **the velocity of actual settlement is generally lower than the Hazen velocity**.

To limit the influence of these phenomena, one perforce must try to achieve a laminar flow in the basin and to reduce the perturbations due to the non-homogeneity of flow by placing obstructions or partitions or making a zigzag channel.

LAMELLAR SETTLERS

They work on the **principle of enlargement of the area of settling surface,** resulting thereby in:

— either improvement in the efficiency of removal (for a given discharge, the desired lowest velocity of fall of particles to be intercepted is reduced by enlargement of the surface area of settling);
— or augmentation of the admissible maximum discharge;
— or reduction of the size (or area) of the work (in comparison with a classical settle).

For continuous sliding and removal of sludge, a steep inclination (50 to 60°) of the trays, tubes or cells is required.

For the lamellar settlers various types of **trays and tubes** are possible: trays of plane or corrugated parallel plates and stacks of square, circular or hexagonal tubes etc. Cellular blocks of varied shape can also be used. Each is characterised by a certain projected horizontal area available for settling per m² of horizontal area of the tank.

Three types of settling tanks can be distinguished with respect to direction of sludge movement:

— counter-current (sludge and water flow in opposite directions),
— parallel flow (sludge and water flow in the same direction),
— cross-current (sludge and water flow in directions perpendicular to each other).

At present, the cross-current system is apparently the best adapted to limiting the re-entrainment of deposited particles.

The principle of separation of SS and floatable oils in a cross-current sludge-flow device with two trays placed one above the other in a circular tank is illustrated in Fig. II.9.

Lamellar settlers can be used for separation of suspended matter as well as flotsom such as oil and grease.

For the latter, a German code (DIN 1999) defines the procedure to be used for testing these devices. Measurements are based on a synthetic effluent composed exclusively of mineral oils (5 g/l) and water. This has nothing

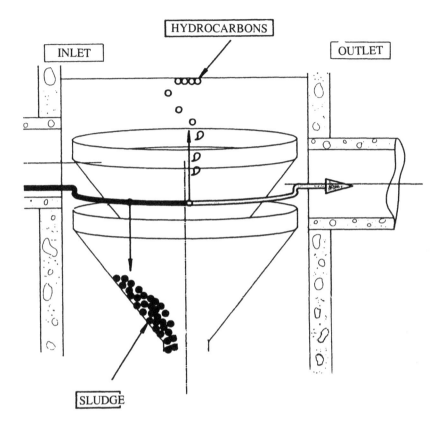

Fig. II.9 Separation of SS and hydrocarbons in a cross-flow lamellar settler (ITERA document).

in common with SS. The results claimed by some manufacturers (95% retention of mineral oils) cannot be extrapolated to removal of suspended matter from stormwater.

The conditions necessary for obtaining good efficiency depend on the criteria for dimensioning, which differ widely depending on whether it is desired to intercept SS or oil and grease. For SS, a Hazen velocity of around 1 m/h is necessary to obtain an efficiency of removal of the order of 80%; for hydrocarbons velocity is about 8 m/h. The velocity and settling time adequate for retention of SS generally suffice for efficient interception of oil and grease also.

Let us recall that the major part of oil and grease in stormwater is adsorbed on suspended solids. In these conditions, the primary objective should be interception of SS. When designing lamellar settlers at the outlet of a retention basin to serve as the finishing treatment, one should take into account the settling obtained in the body of the basin itself. For the lamellar settlers to be efficient, they must be dimensioned for Hazen velocities smaller than those obtained in the upstream basin. This arrangement can be justified only in the case of very sensitive receiving waters.

The few experiments conducted on lamellar settlers have shown that the average removal efficiency for oil and grease is of the order of 50%. Efficiency with SS is highly variable depending on the intensity of the storm. It may vary from 0 to 84% for settling/rising velocities between 0.2 and 12.6 m/h in a device dimensioned for 12.8 m/h (RUPERD [70]).

Generally, the primary condition for obtaining good efficiency is that these devices be cleaned regularly; negative results have often been observed due to the re-entrainment of deposited sludge.

Lastly, efficiency decreases with higher flow rates or if the water is only slightly polluted. It may be added that to obtain good performance of **lamellar settler blocks**, a device should be placed upstream to regulate the flow rate and to ensure proper distribution-diffusion of water. This feature is illustrated in Fig. II.10 showing a horizontal-flow lamellar-block settler.

II-2.2.6 *Other processes*

Other processes are rarely used in France, especially those involving the vortex effect, centrifugal forces etc., such as the static cyclonic separators. These devices are quite complex and involve a high cost/efficiency ratio.

II-2.2.7 *Conclusion*

Table II.5 summarises the type of device for removal of different forms of pollutants as well as the main role of each device. Besides their role vis-à-vis chronic pollution, these devices can, during the dry season, play a protective role against accidental pollutions. Of course, this requires provision of safety valves at the ends of the device to isolate it when necessary.

Although the number of available experimental studies is still small, the units for treatment of stormwater show significant efficiency, which promotes protection of the receiving waters. Their installation upstream of a retention basin is particularly useful when the basin serves as a spot of leisure or pleasant ambience.

In all cases, robustness and sturdiness are the properties to be considered while designing these devices which should be capable of functioning over a long period.

Consideration of objectives of quality of the receiving waters may require dimensioning the pretreatment units for a flow rate corresponding to storm episodes with a periodicity of four times in a year. A weir and a conduit for bypass should be provided for heavier rainfalls.

Lastly, as in the case of an urban or industrial effluent treatment work, **the maintenance and upkeep of pretreatment works should also be properly planned**; in fact, the efficiency of these works is highly dependent on it.

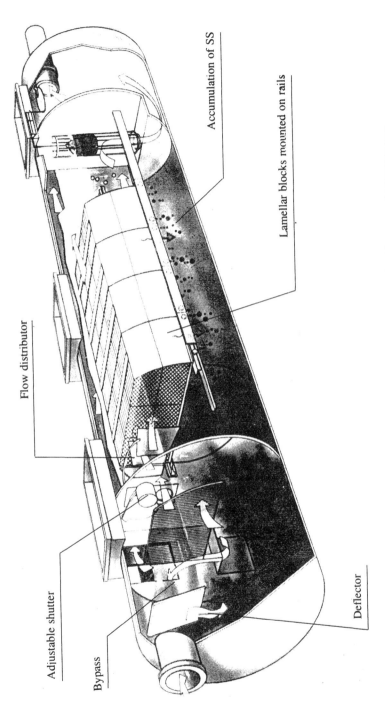

Flow distributor

Adjustable shutter

Bypass

Accumulation of SS

Lamellar blocks mounted on rails

Deflector

Fig. II.10   Horizontal lamellar settling tank with mobile lamellar blocks. (document SAINT-DIZIER)

Table II.5   Pretreatment units: types and roles

Device	Operation/Type	Type of pollution eliminated	Role of device
Screens	Manual Mechanical	Macrowastes (8 to 80 mm)	Protection of downstream units Rendering the flow 'aesthetic'
Grit chambers	Longitudinal grit chamber	Particles larger than 200 µm	Trapping of gravel, sand and grit
	Longitudinal grit chamber 'with constant velocity arrangement'		Protection of equipment (pumps)
Settling tanks	Normal or simple settling tanks	Particles larger than 20 µm	Reduction of suspended solids and absorbed/adsorbed heavy metals, organics and hydro-carbons.
	Lamellar settlers		
Oil and grease removers	With flow-breakers (connectors)	Oils, greases, flotsam	Rendering the flow 'aesthetic'
	Prefabricated cleanser-deoiler		Trapping floating accidental pollution

Lack of maintenance can render them unusable—even harmful for the natural environment—in certain cases when re-entrainment of trapped particles occurs. It is advisable that routine maintenance be carried out at least once in six months in the case of pretreatment works for retention basins and after every storm episode for protection of screens. Local experience of the management of drainage networks can help in deciding the frequency and nature of routine maintenance.

II-2.3   Outlet or Evacuation Works

Evacuation works, also called outlet works, play an important role in the operation of a stormwater basin. Depending on the type of basin, these works are meant to:

— control the water level during dry or low-water periods,
— maintain a desired rate of outflow,
— safely evacuate excessive water flows, should rainfall be heavier than the design capacity of the basin,
— connect with the drainage channel,
— empty the basin through the bottom,
— protect the hydraulic installations (screens, flow-breakers, settling tanks, oil and grease removal units etc.) and/or negate pollution of the receiver waters.

Depending on the downstream constraints, the design of an evacuation work may vary considerably. A few types are listed below in the order of increasing complexity:

— Simple nozzles delivering a discharge dependent on the upstream water head; this device is widely used.

— Regulators with constant discharge; various types of this device will be described later.

— Regulators with variable discharge used for automatic control. This type of control is presently in the preliminary stages of development. Most of the adjustable valves are controlled mechanically or electrically.

Various devices may or may not be integrated in one and the same work. The characteristics of the ensemble shall depend on:

— the site;

— the local topography and, the head available for regulation of discharge depending on whether the basin is built in plane areas or in uplands;

— the discharge per se;

— the freeboard or variation in water head over which the discharge is to be controlled;

— downstream hydraulic capacity.

The number of parameters to be taken into account is thus quite large. It is very rare that two identical systems may be optimal for two different water bodies even for fulfilling the same objectives.

II-2.3.1 *Control of water level during low-water periods*

This objective is usually achieved with a weir crest adjusted to the minimum desired water level in the case of a detention pond. For quick attainment of the desired flow rate and for taking full advantage of the available evacuation capacity, **the weir should have a large developed length**. Such weirs allow passage of discharge under a small head. Thus the Giraudet toothed weirs, or those called 'duck's beak', were very widely used on small rivers for regulating the water depth in upstream ponds. A variation of just 10 to 15 cm in the water head (the ten-year frequency freeboard is 50 cm) should help maintain the regulated discharge. A zigzag or toothed geometrical shape helps to obtain a large weir length in a civil work of reasonable dimensions. The pit downstream of the weir should be able to withstand the pressure due to the momentum of falling water and hence should be ballasted.

The formulae for calculating the discharge in m^3/s per unit running length of weir (free flow) are of the form:

$$Q = a \times \sqrt{2g} \times h^{3/2}$$

The coefficient 'a' varies from 0.34 to 0.40 depending on the type of weir; g is the acceleration due to gravity ($g = 9.81$ m/s^2); h is the water depth above the weir, expressed in metres.

Photo 6   Zigzag or toothed weir for large developed length.

II-2.3.2   *Regulation of discharge*

There are numerous systems for control of discharge, varying from the simplest to the most complex. These are often hydraulic devices controlled by the upstream and downstream water levels. Some systems were originally designed for controlling flow in large irrigation networks operating on gravity. More recently, devices have been designed specially for regulation of flow in drainage and sanitary systems.

CALIBRATED ORIFICES

These are simple devices. Their performance is satisfactory for controlling the discharge when the difference in water depth corresponding to zero discharge and maximum discharge is small compared to the initial depth. They can be used when an accurate regulation of discharge is not a requirement. The discharge Q in m^3/s measured by such orifices (case of submerged orifice) is given by the formula:

$$Q = m \times S \times \sqrt{2gh}$$

where m is a coefficient of contraction, always < 1, often around 0.6; h is the water head above the axis of the orifice, in m; and S the area of cross section in m^2.

Adequate protection by provision of a screen is a must to preclude obstruction of the opening.

FLAP MODULES

Flap modules of small capacity comprise a set of valves which are opened until the desired value of discharge is attained, the discharge being regulated litre by litre, or ten litres or more per second.

The exact discharge delivered varies between 95 and 105% of the nominal discharge for a given variation of the upstream water depth, above and below an optimum level. Charts supplied with each module give an accurate idea of its performance. The simple small modules cannot operate with a large level difference upstream—a valve with constant downstream level has to be provided with it. This valve can stand large variations in the upstream level of the water body at the time of rainfall and maintains, on the downstream side, small variations of water level which are within the capacity of the flap module.

For large flow rates, a module with double flaps that can withstand large variations of upstream water level has to be used. Often, the variation of level is compatible with the difference between the level during flooding and the nominal level, and complementary regulation by means of a valve to constant downstream level is not necessary. Figs. II.11 to II.13 illustrate the functioning of modules with single and double flaps, designed by ALSTHOM FLUIDES SAPAG (earlier known as NEYRTEC).

The position of the module should be so adjusted that the nominal level in the basin corresponds to a level 10% below the set flow (Q – 10%) of the module.

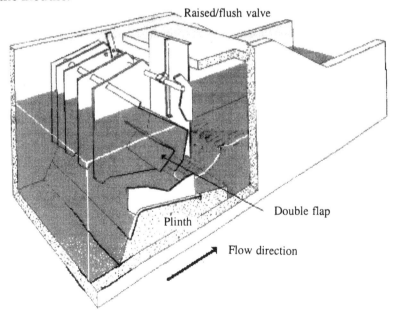

Fig. II.11 Perspective view of a flap module (ALSTHOM FLUIDES SAPAG document).

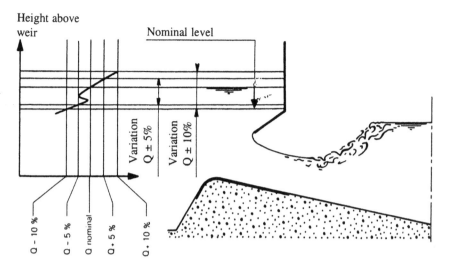

Fig. II.12 Module with single flap and its performance curve (ALSTHOM FLUIDES SAPAG document).

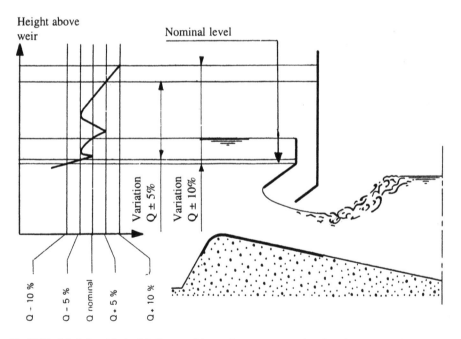

Fig. II.13 Module with double flaps and its performance curve (ALSTHOM FLUIDES SAPAG document).

VALVES WITH CONSTANT DOWNSTREAM LEVEL

These valves are essential components of the devices for regulation of discharge and water depth in large irrigation networks working on gravity. In France, they are fabricated mostly by Alsthom Fluides Sapag under the brand names AVIO and AVIS valves. Their sturdiness makes them ideal for use in drainage works. Valves manufactured by Lezier may also be mentioned in this context.

A valve of this type is placed between the weir and the device to control the discharge (flap module, overflow weir calibrated orifice etc.). In the context of a stormwater basin, it is not necessary to use a grit chamber upstream of the valve as settling will have taken place in the basin itself. On the other hand, for the evacuation of water stored temporarily in a dry basin, it is almost obligatory to place a grit-removal device upstream of the valve.

There are two types of valves with shutters in the form of curved plates controlling a float. Only the AVIO valves are useable for regulation of the discharge of outflow from stormwater basins. In fact, the AVIO valves operate for large head of water while the AVIS valves are particularly appropriate for small flow rates.

An AVIO valve is an oscillating device which has in front (i.e., directed upstream) a flap valve working as a shutter and in the rear, a float. The float follows the variation of the water level downstream of the valve. If this level rises, the float is raised and causes pivoting of the flap which descends, thereby reducing the discharge delivered and vice versa. The principle of functioning is illustrated in Fig. II.14.

Valve choice depends on the regulated discharge and maximum water depth when the rate of outflow is equal to that shown in Fig. II.15.

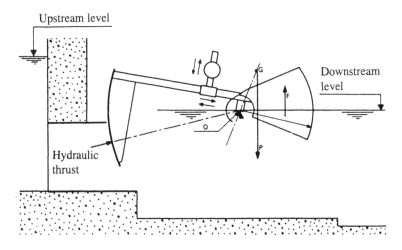

Fig. II.14   Profile of an AVIO valve (ALSTHOM FLUIDES SAPAG document).

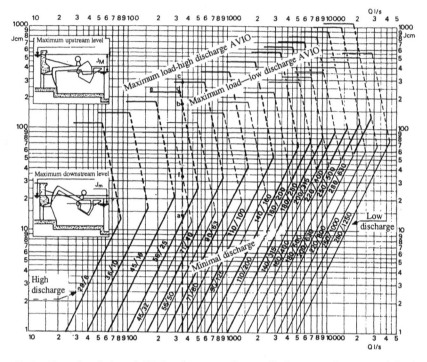

Fig. II.15 Chart for choice of AVIO valves according to discharge, with maximum load at zero discharge and admissible rise in water level corresponding to a given variation of discharge around a central value (ALSTHOM FLUIDES SAPAG document).

There are other types of valves whose principle of working is slightly different (cylindrical valves etc.). These valves are followed by a simple weir whose width, fixed once for all, controls the discharge and the accuracy attained is reasonably good. To attain a higher accuracy of control of discharge, a set of modules with a flap of the type described above must be used.

ADJUSTABLE SHUTTERS WITH FLOAT

The rate of outflow is controlled by a shutter which closes the orifice of evacuation to a degree directly varying with the rise of water level in the basin; thus the rate of outflow is practically constant.

In the course of rise of water, displacement of the shutter in front of the opening is controlled by means of cams or gears activated by an arm to which the float is attached at the tip (Fig. II.16).

These devices should be well protected against foreign bodies (dead leaves, twigs, plastics etc.) which can block the mechanisms.

Manufacturers are numerous. In France, ISD Environnement (brand name HYDROSLIDE) and in Germany, STEINHARDT, FRG.

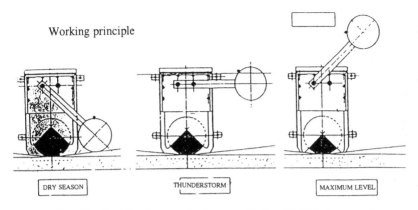

Fig. II.16 Sketch of functioning of a shutter with float (ITERA document).

FLOATING WEIRS

These evacuation devices maintain practically constant discharge. Water inflow takes place at the surface. A weir with overflow to a hopper is carried by two lateral floats designed for a particular range of discharge. Water overflows into a rigid or flexible conduit attached to the hopper. This type of arrangement is often adopted when the outlet system includes a device for separation of hydrocarbons.

OTHER TYPES OF CONTROL DEVICES

This category includes:
— regulators with chamber of rotation of the 'Hydrobrake' type, which work on the principle of outflow;
— regulators with vortex (SEPAREPUR BROMBACH);
— regulators with membranes of the 'Flow Valve' type in which the discharge is controlled by a venturi whose opening is modified by the pressure of water on a capsule.
These devices are seldom used in France.

RELATIVE DEVICES WITH VARIABLE FLOW RATES

The outflow from a basin can be modulated as a function of one or several parameters upstream of the basin (inflow rate or water depth) and/or downstream conditions (type of possible utilisation of outflow, state of connected drainage, extent of various deposits etc. (Fig. II.17).

The degree of opening of valves, or their state (all or nothing) or cross-section of adjustable shutters are controlled by parameters determined by automation or other devices of command. In particular, systems of automatic operation can be linked to computers or microprocessors, either located on site or in a central control room. Equipment for real time control (RTC) are generally included in such systems. Valves with sections, flaps, slides,

Photo 7   Shutter with 'Hydroslide' float (ISD Environnement document)

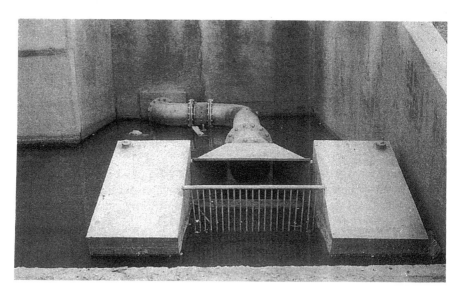

Photo 8   Floating weir (ISD Environnement document).

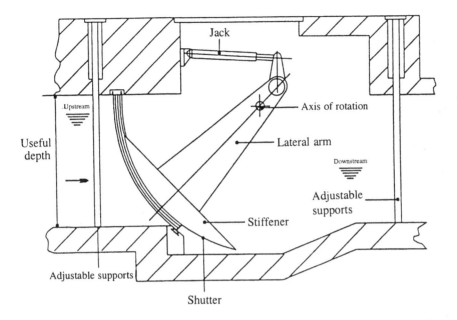

Fig. II.17   Example of valves which can be controlled by water depth or discharge.

guides or lids can be controlled by this type of command systems. Such devices can be used to optimise protection against floods as well as against pollution.

EVACUATION PUMPS
It may be recalled that when evacuation by gravity is not possible (general case of underground tanks) recourse is taken to pumps, submerged or placed on the surface, and controlled by the water level in the basin. Also, they are well adapted for very low flow rates (1 to 10 l/s) or for delayed emptying (zero discharge during rainfall for basins arranged in parallel or as a network).

II-2.3.3   *Weir for excessive floods*

When a basin is dug in a natural terrain, its overflow cannot cause serious problems. At worst, when it is overfull, the stormwater from the terrain drained by it may not be able to drain off by gravity and the environs may remain flooded for some time. In contrast, when the basin is bounded by a dyke, its submersion may entail rupture of the dyke and almost instantaneous emptying of the pond, causing considerable damage downstream. A spillway is a device that protects the dyke; it is indispensable in the case of basins with an earthen dyke or embankments and should be adequately dimensioned.

There are two distinct types of spillways for evacuating floodwater:

SURFACE SPILLWAYS

These often involve provision of a special spillway section in the dyke, followed by an inverted bucket terminated by a hydraulic jump for dissipation of drop velocity. The bucket flooring and the nearby downstream embankment should be particularly strong and the dimensions large enough to preclude endangering safety of the work.

For small ponds, a lateral spillage channel can be used, which is dug in the natural terrain and is totally independent of the dyke.

Barring exceptions, safety siphons (of the type Alsthom Fluides Sapag or Lezier) are not used for stormwater basins. Nevertheless, if space is lacking for construction of a standard lateral spillway channel, this type of device may be resorted to. The principle of working is the same as that of common siphons, as explained in Fig. II.18. Completely primed, the discharge of the siphon is almost equivalent to that of an orifice of the same transverse section placed at one side of the drain, i.e., at a much lower level than a spillway weir. The discharge is thus larger than what would pass over the same width weir on the surface. For very large discharges a single siphon or set of siphons placed side by side may be used.

TUNNEL SPILLWAYS

This type of spillway is constructed in accordance with norms applicable in the case of dams. Here water is transported downstream of the dyke by a large-size tunnel passing under the dyke fed by a well or a tower equipped with a circular tulip-type weir [54].

II-2.3.4 Bottom-level evacuation

In the case of detention pond, a device for evacuation of water at the bottom must be provided. For a pond bounded by a dyke the device works by gravity and should be included in the design. In the case of a basin at the bottom of a pan, evacuation is accomplished by a mobile pump with easy access.

Evacuation allows inspection of underwater banks and structures as well as any alterations or additions, e.g., silt removal or control and regulation of fish population (routine operation in fish ponds). The rule of cleaning once every ten years applied to dams and large reservoirs is recommended for stormwater basins also. If the water quality is degraded, it must be evacuated after appropriate treatment during a period when the receiver water is less sensitive. The basin gets subsequently filled with good quality water.

Dimensioning of the evacuation conduit should take the following factors into account:

— desired flow rate or rate of inflow of groundwater; it is known that evacuation is preferably carried out during cold humid weather;

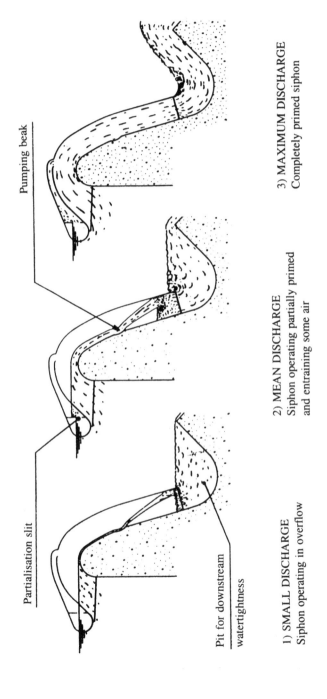

Fig. II.18   Three stages of operation of a safety evacuation siphon. (documents ALSTHOM FLUIDES SAPAG)

— rate of evacuation, so calculated that the pond is completely emptied in at most ten days.

When a dyke is to be constructed across a stream and a diversion conduit built to carry the flow during execution of the work, the same conduit may serve for future bottom evacuation also.

II-2.3.5 *Protection against quicksand*

The durability of dykes can be seriously endanged by the formation of cracks, tunnels, holes around the water outlet from the basin; such tunnels or holes can be caused by various factors (rodents, roots, unauthorised human traffic, seismic vibration, hydraulic or mechanical compression etc.) and the existence of conduits across a dyke can be a major catalyst for formation of such defects. Great care has therefore to be exercised while laying conduits in or under a dyke that closes a basin, viz:

— conduits carrying flood spillage,
— conduits for bottom evacuation.

Such conduits should be equipped with antiburrow devices. The earth mantle around the conduit should be firm and compaction done very carefully. In all cases, adequate attention should be paid to the quality of materials and its contiguity with the conduit. Placing antiburrow flanges or rings radially around the conduit at 0.3 to 0.8 metres is an excellent technique (Fig. II.19). These should be particularly emplaced in impermeable zones (core) where hydraulic gradients of infiltration are steepest (BERGUE [3]).

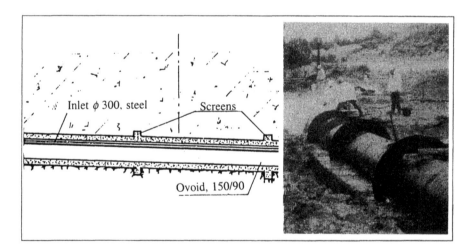

Fig. II.19   Part of a dyke showing a conduit equipped with antiburrow rings [54].

II-2.3.6 *Junction with natural drain of evacuation*

The conduit joining with the natural drain of evacuation may carry regulated releases of flood spillages, flows from evacuation at the bottom or various combinations of these flows. The number of conduits crossing the dykes should be kept to a minimum by regrouping the inlet sizes if necessary. These conduits should be accessible to easy maintenance and control.

II-2.3.7 *Description of some typical outlet works*

Figures II.20 to II.23 show common types of outlet works in which several functions discussed above have been regrouped.

Outlet works of type 1
AVIO valve + module with flaps, all kinds of discharges
(maximum 30 m³/s per unit)

The objective of this type of work is to regulate all kinds of discharges to the extent for which there is sufficient space for installing necessary mechanical devices. It is integrated in the embankment or the dyke and comprises:

— overflow section of the dyke protected by appropriate screens;
— connecting channel, also equipped with a screen for protection. It carries water towards the control devices (AVIO valve and module with flaps);
— AVIO valve ensuring a constant level downstream. It requires installation of a stilling chamber, the size of which has to be enlarged for larger flow rates. A primary guard valve isolates the unit for maintenance of the AVIO valve;
— module with flap installed at the exit of the stilling chamber. This apparatus protrudes into the conduit or the evacuation channel. It can be replaced by a simple calibrated (but less precise) weir;
— evacuation conduit at the bottom, downstream of the flow-control device and normally operated through a sluice valve.

Outlet works of type 2
Module with double flaps, high flow rate
(model CC2 = 200 l/s per unit or more)

The objective of this work is to regulate the discharge of several cubic metres per second. The only limitation is the space available for eventually placing several modules side by side.

This device uses modules with double flaps and can operate even when the variation of upstream levels is much larger than that at which the modules with single flap can operate. These modules can function without AVIO valves if the maximum freeboard is smaller than the variation of both levels

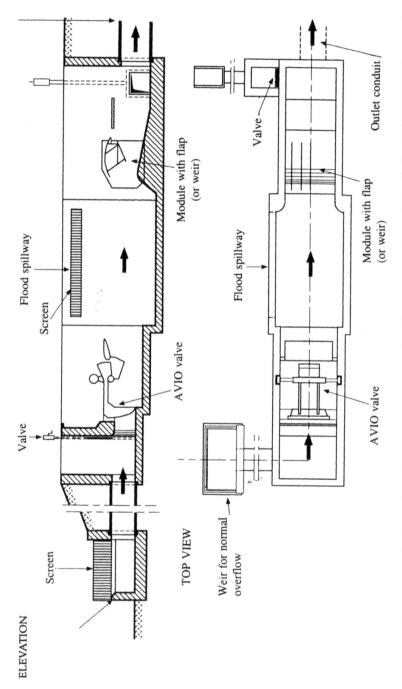

Fig. II.20 Outlet works of type 1 with a valve maintaining constant level downstream and a module with flap (SAUVETERRE document).

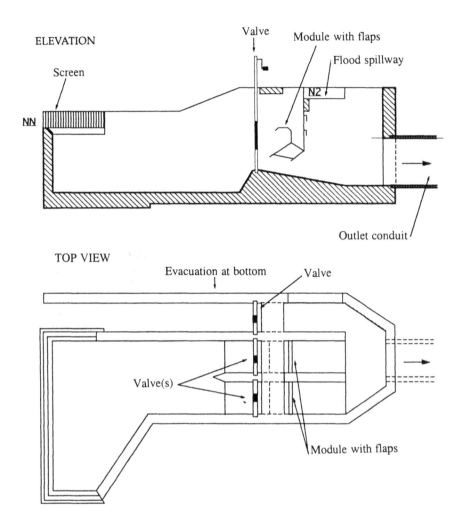

Fig. II.21   Outlet works of type 2 with module with double flaps (SAUVETERRE document).

admissible for them (see manufacturer's catalogue). It may be noted that the freeboard to be taken into account is the one corresponding to the basin filled to the spillway crest if there is one and not just that for the observed occurrence of flood at the project. For larger rise of water levels, the safety weir is activated and water can flow above the flap module also.

Outlet works of type 2 comprise:

— overflow weir of large developed length with protective screen,
— module or modules with double flaps placed on a concrete structure,
— spillway crest constructed in lateral wall of channels,

— evacuation at the bottom through a lateral pipe, the flow being carried into an evacuation conduit; it is generally closed by a valve.
The normal layout of the work is perpendicular to the bank.

Outlet works of type 3
Calibrated orifice, low discharge, 50 l/s to a few hundred litres per second

---

In addition to the screen placed in front of the overflow weir, another screen should be placed in front of the calibrated orifice itself.

The flood spillway and evacuation at the bottom can be readily incorporated in the work (Fig. II.22).

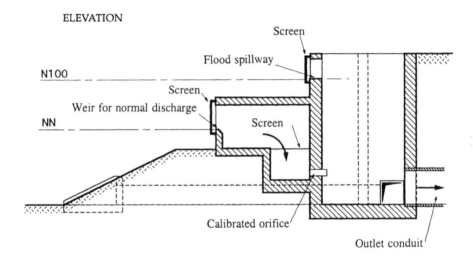

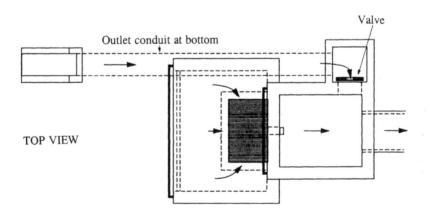

Fig. II.22   Outlet works of type 3 with calibrated orifice (SAUVETERRE document).

Outlet works of type 4
Adjustable shutter with float, flow rate between 7 l/s and 1.5 m³/s

During the rise of water in the basin, the sectional area of the outlet orifice is controlled by means of a float and a cam. In this case, too, protection through screens is essential; also, free movement of the rod of the float must be carefully controlled (Fig. II.23).

SECTION

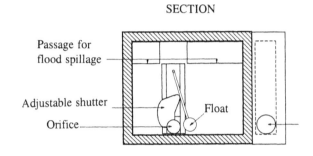

ELEVATION

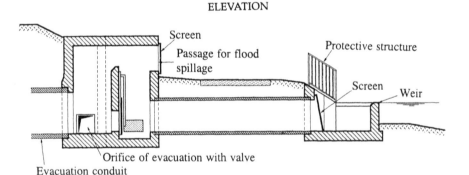

Fig. II.23   Outlet works of type 4 with adjustable shutter (SAUVETERRE document).

Table II.6   Summary of controlled evacuation devices

Device	Flow rate	Accuracy of regulation
Valve maintaining constant level downstream	25 l/s to 30 m³ /s	good (*)
Module with flaps	5 l/s to unlimited	excellent (± 10%)
Fixed calibrated orifice	5 l/s to a few m³/s	moderate
Adjustable shutter with float	7 l/s to 1.5 m³/s	excellent
Floating weir	10 to 150 l/s	excellent
Control device	unlimited	excellent

*If attached to modules with flaps or a calibrated orifice.

Table II.6 gives a list of controlled evacuation devices from which a choice can be made depending on the:
— order of magnitude of regulated flow rate,
— accuracy desired.

II-3   OPEN BASINS: OTHER REQUIREMENTS

II-3.1   Banks and Embankments

II-3.1.1   Banks of water basins

Except for the very cramped sites in which the area up to the plummet of the water body is constricted, or for architectural considerations, the banks called 'natural' are generally erected in the following manner, going from the exterior towards the basin:
— An embankment of small slope, linking up with the natural terrain and covered by turfed or planted earth. A slope (height/base) of 1/3 is the minimum compatible with mechanised maintenance, including mowing of lawns.
— A horizontal platform of variable width for areas of relaxation and paths for promenade or maintenance, between 0.5 and 1 m above the water body, turfed or covered and sloping towards the water bank; this space gets inundated very rarely (less than once in ten years).
— Embankment of slope 1/2 to 1/3 or a vertical facing delimiting the water surface; this somewhat steep slope reduces the disagreeable looks of the freeboard corresponding to the normal operation of the basin. The embankment is generally protected against erosion by covering it with dense vegetation, or with pitching slabs, logs, rocks, or geotextile membranes with a fishnet type structure permitting plant growth.
— On the zones accessible to the public, a horizontal berm for safety, at least 1.5 metres wide wedged at a depth of around 50 centimetres below the permanent water level.
— An embankment below water of average slope between 1/3 and 1/5, depending on the geotechnical characteristics of the material constituting the embankment. Locally, this slope is further reduced if growth of rooted aquatic vegetation is to be encouraged.

For a typical profile of a natural bank the reader is referred to Fig. I-12 in Section I-6.4.1.

It is always advisable to choose options which allow on-site reutilisation of mud or earth dug out for making flower-beds, noise barriers and artificial relief. Transport of mud away from the site is always an expensive operation.

II-3.1.2   Techniques of protection against wind erosion

Experience has shown that a bank exposed to wind can shrink by several metres in a few years. The techniques commonly used for ensuring protection of banks against erosion either by plant cover or by different forms of pitching, are summarised in Table II.7.

Table II.7   Techniques of protection of banks against erosion

Type of protection	Advantages	Shortcomings
Turfing	Less expensive	Frequent maintenance necessary
Small shrubs	Can be used for steep slopes	Cost dependent on species (70 to 150 F/m^2, compost inclusive)
Aquatic plants	Possible use of biomass produced Good trapping of flotsam	Cost 100 to 150 F/m^2
Wire mesh, Enkamar brand or equivalent	Rot-resistent: retains flotsam, allows vegetation	Extra geotextile and ballasting under water required
Alveolar slabs, Evergreen brand or equivalent	Turfing easy Excellent protection	Not aesthetic unless good growth of grass takes place Cost 200 to 400 F/m^2 with compost
Steps either paved or made of pebbles	Excellent protection Good aesthetics	Cost high
Rocks, blocks 30 · 50cm, height 1 m with geotextile	Excellent protection	Cost more than 500 F/m^2 (1993, without taxes)
Concrete wall, height 1.5 to 2.5 m	Constitution of berms for promenade on water banks possible	Cost high Foundations required Surface has to be treated to give appearance of stones
Metallic plates	Same as above	Facing indispensable Cost high
Stabilisation with rot-resistant wood: — facings made of logs — grooved planks or — supported plates	Excellent aesthetics	Not suitable for height greater than 1 m

II-3.1.3   Embankments of dry basins.

Strictly speaking, dry basins do not require embankments as they are generally totally enclosed by natural banks.

Construction of embankments, if provided, is subject to the same rules as those relating to the detention ponds but a safety platform is not needed since access to these basins is, in principle, forbidden during the storm events.

Traditionally, a straight or sinuous pan is provided at the bottom of the basin (see photo 9) which carries the final stage of the evacuation of

Photo 9   Pan at bottom of dry basin; Griffon à Vitrolles park (Bouches-du-Rhônes), France.

stormwater and drainage of eventual flows during the dry period (see Sections on sullage, cross-stream flows). This pan can also expedite removal of sediments and wastes. It is usually cemented or lined with stone slabs or tiles for aesthetic appeal.

Bank slopes are designed with due consideration for their stability as well as compatibility with pedestrian movement and mechanical maintenance.

II-3.2   Access

Let us recall that although it is at the expense of valuable space, the decision makers are obliged to provide access to basins for maintenance machinery (boats, trucks etc.) as well as space for corresponding manoeuvres. This is positively a major burden for small retention basins for which the access space requirements constitute a significant part of the total space.

In urban and suburban regions, acts of vandalism have become a constraint to be taken into account during the design stage. All structures necessary for the operation of basins (devises of regulation pretreatment) should be protected by impenetrable fencing and access should be restricted exclusively to maintenance personnel.

II-3.3   Vegetation

The first rule to be kept in mind while planning a basin for stormwater, dry or wet, is that it should conform to the rules of nature. For planting and maintenance of vegetation, the following principles must be adhered to:

— never plant trees on a dyke;
— never plan development of a plant species which is not adapted to water depth and nature of the soil. Also, avoid plants which can proliferate and also mono-specific plants which generally do not conform to the rules of nature;
— establish and maintain the biodiversity of the local vegetation;
— grow plants of natural species of helophytes on banks, such as those seen along natural ponds in the environs. Plant mixed species with same needs [87];
— for decorative effects, clusters of monospecific horticultural species can be created. In this case one has to eliminate or limit the natural species on the banks as well as in water because they tend to damage horticultural species. Furthermore, some decorative species do not grow spontaneously in a new environment, for example the waterlily. They can be introduced only if one has very good knowledge of the species and their requirements regarding water depth;
— landscaping should be so planned as to allow access and movement of machinery for maintenance of the planted area.

It is an accepted fact that natural vegetation is never permanent; a succession of plant groups will appear over time after the installation of a stormwater basin.

II-3.3.1   *Case of detention pond*

At the early stage, filamentous algae are plentiful and may provide abundant bloom (see coloured photos 15 and 16). They disappear on their own. Two or three years after the basin is commissioned the ubiquitous pioneer species (for example *Typha latifolia*) develop and a dozen or so species of diversified reeds appear. Human intervention can result in hastening this normal evolution towards a 'climax' of natural ponds and be helpful in maintaining it.

In open water, hydrophytes flower on the surface are accused of tarnishing the mirror effect of the basin. The public should be informed that this is not a pollution but a natural phenomenon. Nevertheless, helophytes with aerial leaves should be eliminated in open water zones where the mirror effect is desired. In the early stages of growth, these plants can be easily removed. If they are allowed to root and develop, eradication becomes difficult.

II-3.3.2   *Case of dry basins*

The distribution of plants should take into account the duration and level of submersion and the situation regarding the groundwater level during different seasons. If the groundwater level changes significantly following an increase in withdrawal of water (for drinking water supply, irrigation etc.)

or blockage of phreatic flow at the time of construction of permeable trenches (for roads or railway lines, etc.) an evolution of planted species may be observed. These species will wither and be replaced by other better adapted ones, either spontaneously or by human intervention following diagnosis of the state of the vegetation.

II-4   UNDERGROUND TANKS

The construction of underground tanks is subject to geotechnical constraints of the same kind as those applicable to open basins. The civil works in this case are generally large in size. Information relevant to soil and earthwork given in Sec. II.1 is also applicable to underground tanks. Also, in this case the environment of the workyards is more constraining because they are located in dense urban surroundings. After construction, modifications and repair of works is often difficult. Hence it is imperative that right at the design stage various ancillary structures (access, repair yards on the premises, maintenance facilities) be taken into account and the construction carried out with utmost care.

II-4.1   Construction Aspects

II-4.1.1   *Influence of hydrogeology and geology on construction*

The presence of groundwater poses various problems of execution and constitutes an important factor in the overall plan of the project. Management of the water table during construction and during the life of the structure is one of the major constraints with respect to the selection and follow-up actions for an underground tank. The following factors need to be considered:

— Groundwater may infiltrate into the basin. It may be tolerated to a certain extent but the flow rate of infiltration must be limited.
— Hydraulic pressure on the walls cannot be ignored and may impose the restriction that the tank be cylindrical in shape or anchorage provided.
— The upthrust exerted by the groundwater under the floor of the tank tends to lift it; various solutions are discussed later to remedy this difficulty.

The earthwork must be executed with utmost precaution and, if necessary, the water table may have to be lowered by pumping.

CONSEQUENCES OF LOWERING THE WATER TABLE

Lowering the water table by one or several metres requires accurate study of the pumping procedure. It must be executed by a team of specialists in hydrogeology. The following are the risks involved in this operation:

— entrainment of fine particles during pumping, creating voids and thereby possible cave-ins of adjacent ground;

— reduction in soil bearing capacity. In the case of muddy, clayey or peaty soils, reduction in water content implies a large reduction in volume, which can lead to shrinkage and constitute **grave risk for the stability of buildings situated in the vicinity**. Shrinkage depends on the nature of the subsoil, initial load on the surface and the amount of water reduction. Pumping exerts an influence in some cases over several hundred metres from the point of pumping.

Implementation of a system of pumping thus requires careful study. It should be carried out only to the extent that the constructed structure can bear the extra pressure generated.

INFLUENCE OF CONSTRUCTION ASPECTS

Various techniques are available for limiting high pressures; they are described in manuals such as DTU* 11.1. The choice of solutions depends on the nature of the soil and the requirement for regular pumping throughout the life of the structure. For example, in Seine-Saint-Denis the pumping required is of the order of 30 m³/h.

When the soil is of good quality and the water table located rather low so that excessive pressure is not generated, and it is possible to excavate up to sufficient depth, the following procedure may be adopted (Fig. II.24):

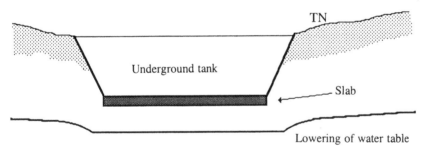

Fig. II. 24 Banking an underground tank.

— after lowering the water table, the zone is banked using embankments as walls,
— the floor of the tank is installed,
— the walls of the tank are erected,
— the perimeter is banked.

In this case, the floor is designed as a slab. Its thickness is the same as that of an industrial slab and varies from 15 to 20 centimetres.

---

*Document Technique Unité: Technical document on urbanisation prepared by the French Ministry of Public Works

When the soil contains several aquifers, a different procedure is adopted. If the groundwater underlying the tank is under pressure, either its effect must be nullified or the pressure countered by an opposing force. The choice depends on several criteria: cost, space available, geotechnical characteristics of the terrain and residual discharge to be evacuated.

When geological considerations allow, **the walls are lowered to a level below the aquifer or to an impermeable substratum in which the tank is anchored**. This helps in isolating the bottom of the structure from the aquifer, reducing the high pressure and also the residual pumping required. The technique used in this case is that of moulded walls (Fig. II.25). The water contained in the part of the terrain enclosed between the walls is pumped out before proceeding with the work of banking. The lateral watertightness obtained through moulded walls is generally adequate.

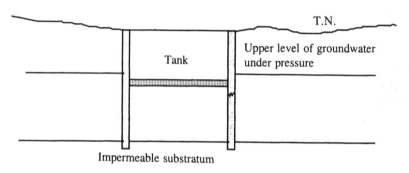

Fig. II.25   Moulded walls for watertightness.

When it is not possible to anchor the walls in an impermeable substratum, three options are available:

i) <u>Creation of a watertight layer:</u> An artificial substratum is created by injecting a silica gel, silicates or a mixture of bentonite and cement in the soil under the basin over a thickness of 3 to 5 metres. The injections seal the natural voids in the soil, thereby creating a watertight bulb (Fig. II.26). The vertical force caused by the groundwater under pressure is buffered by this plug. However, a residual discharge still remains which must be evacuated. Hence a drainage device is provided under the floor. This solution dispenses with the need for lowering the water table during the work.

ii) <u>Constructing a concrete floor:</u> A high upthrust can be compensated by building a thick (between 3 and 5 metres) concrete floor. It is necessary to obviate any chance of water leakage between the wall and the concrete floor (Fig. II.27).

iii) <u>Anchoring the floor:</u> This solution is adopted when the upthrust is not excessive. The floor of 40 to 60 centimetre thickness is anchored in the soil by means of pins or rods. If necessary, it is also anchored in the walls. Pumping of water has to be maintained during the construction work to preclude the quicksand effect (Fig. II.28).

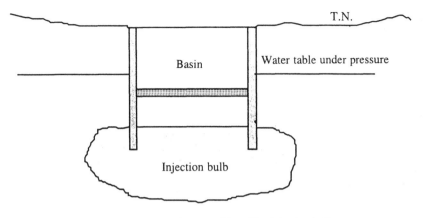

Fig. II.26   Watertightness achieved by injection bulb.

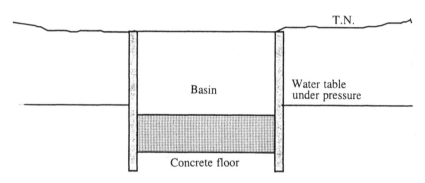

Fig. II.27   Installation of a concrete floor.

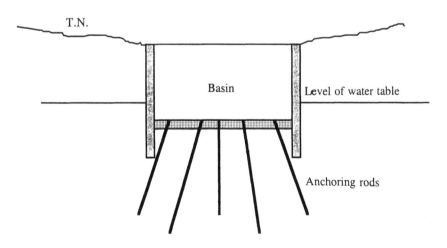

Fig. II.28   Anchoring the floor.

PERMANENT DRAINAGE

As no soil is perfectly impermeable, a permanent system of drainage should be installed. This system evacuates water which rises from the underlying water table. The drainage is installed just under the floor of the basin to prevent the high pressures which might destroy it. The drained water is evacuated by pumping or gravity. Should the device for permanent drainage fail, provision of valves will help the water to flow into the tank. Funnels opening into the tank through orifices equipped with valves can also be used for this purpose; by this arrangement the high pressures are controlled and cleaning of the tank expedited.

Construction constraints and considerations related to the position of the water table are summarised in Table II.8 based on data provided by the General Council of Seine-Saint-Denis (France).

Table II.8   Construction constraints and aspects, as provided by the General Council of Seine-Saint-Denis

	Pressure exerted by intervening water table				
Constraint	Thick concrete floor	Anchored floor	Injected subsoil	Foundation on water-tight layer	Intervening water table not present
Risk of quicksand during construction	X	X			
Lowering of water table necessary during construction	X	X			
Permanent residual drainage			X	X	X
High-pressure valve needed in floor			X	X	X
Order of capacity of evacuation of discharge			30 m³/h	30 m³/h	30 m³/h
Volume of earthwork	large	normal	normal	normal	normal
Thickness of floor	3 to 5 m	0.4 to 0.6 m	0.15 to 0.20 m	0.15 to 0.20 m	0.15 to 0.20 m
Use of rods for anchoring floor (as per requirements of DUP)*		X			
Required soil characteristics		anchoring by rods possible	injectable subsoil	depth of impermeable substratum < 25 m	good hold
Order of construction	walls, floor	walls, floor, anchoring	walls, injection, floor	walls, floor	walls, floor

*Déclaration d'Utilité Publique: a French procedure of holding consultation with the public to determine whether or not the work can be declared as one of public interest.

INFLUENCE OF WATER QUALITY

The choice of materials for construction (sand, aggregates and cement) depends on the corrosive nature of the groundwater; the reader is referred to the strict norms formulated by AFNOR (Association Française de Normalisation: French Association of Normalisation).

II-4.1.2 Tank shape

In addition to the geology and hydrogeology, the geometrical shape of the available plot and what is in the subsoil (e.g. sewer, subway etc.) also influence the shape of the underground tank. The shape should be the simplest possible in order to minimise cost.

The tanks may be compartmentalised so that only a part is exposed to frequent rainfall and thus the constraints of utilisation are reduced.

An underground tank may be cylindrical, parallelepiped or of any arbitrary shape. Nevertheless, the trend appears to favour a cylindrical shape since such basins, within certain limits of diameter, are self-stable, irrespective of the nature of the surrounding terrain. Also, the walls are in compression along the entire periphery, which constitutes an asset for construction in concrete. On the other hand, this shape does not allow optimal use of available space.

Other shapes involve introduction of auxiliary systems to ensure resistance of the walls to thrust exerted by water and soil. The available options include:

— dividing the basin into components working in compression, which stiffens the structure;
— anchoring with steel rods;
— covering slabs acting as stiffener;
— intermediary slabs to provide stiffness.

II-4.2 Auxiliary Works

Underground tanks require auxiliary works and equipment for hydraulic management, maintenance and general upkeep.

II-4.2.1 Workshop and equipment yard

These provide shelter for such equipment as electrical machines, generators, sensors, systems of inspection etc. and should be inaccessible to the general public.

II-4.2.2 Systems of evacuation

Various sets of pumps are required for evacuation of water and sludge.

EVACUATION OF WATER

Water is mostly evacuated by means of pumps which also act as regulators (for the case of evacuation by gravity, there are different regulatory devices; these were described in Section II-2.3).

Designers of installations should keep in mind the hydraulic compatibility with the downstream drainage network and ensure the safety of operation.

In selecting a pump, the following criteria must be considered: duration of evacuation, admissible flow downstream, total height of pumping, electric power available and the nature of effluents. Using this information and the charts supplied by the manufacturers, one chooses the type and number of pumps required.

It may be recalled that the power of a pump is given by the formula

$$P = QH_{mt}\, \rho/367$$

where P is expressed in kW, Q in m^3/h and H_{mt} (total manometric height) in metres; ρ is the efficiency of the pump + motor ensemble.

Before installing pumps it should be ensured that while on the one hand no eddies are generated in the lower part of the basin, on the other the range of action of the pumps is compatible, so that they operate in an identical manner (Fig. II.29).

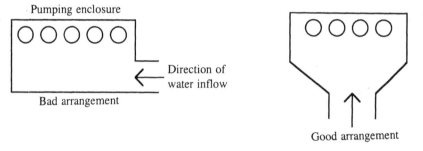

Fig. II.29   Direction of water inflow to the pumping enclosure.

To avoid the risk of stalling, the pumps are generally of the immersed type. Pumps with propeller are sometimes used when there is a large backflow and the pumping height is small.

To obtain the fastest rate of evacuation, one adopts the flow rate equal to the capacity of the downstream drainage network with a set of pumps with adjustable discharge.

For the system to be reliable, provision for one or several spare pumps should be made. In no case should the additional capacity of the set of spare pumps be less than 30% of the required capacity. The pumps should be used in rotation so that no pump is left unused for a long time. Emergency power from diesel generator sets should also be available.

1
Villeneuve d'Ascq (59); one of the earliest stormwater
retention basins constructed in France
(SAUVETERRE photo)

2
Detention pond no. VIII Sud à Marne-la-Vallée (77):
station RER of Lognes-Mandinet (France) spanning the waterbody
(SAUVETERRE photo)

3
Roadside stormwater basin made watertight by canvas sheet
with flow rate regulated by floating weir
(SAINT-DIZIER document)

4
Basin in shape of a canal at Melun-Sénart, France (77),
for which maintenance of water appears difficult
since the dry spell of 1988 to 1991
(SAUVETERRE photo)

5
Dry basin of Liourat à Vitrolles, France (13) located
in a velodrome and football field
(SAUVETERRE photo)

6
Detention pond of Créteil-Préfecture, France (94) located in
dense urban environment
(SAUVETERRE photo)

7
Dry basin of Chemin de Clères, near Rouen, France (76),
of intricate geometrical shape with embankment and
floor made into lawns
(SAUVETERRE photo)

8
Pontoons for fishing in detention pond of Porcheville, France (78)
(SAUVETERRE photo)

9
Detention ponds of ZAC de Bourgenay, France (85) widely used for
leisure activities: fishing, initiation to sailing ('optimists')
and safe pleasure trips in electric 'mini-tugboats'
for small children
(SAUVETERRE photo)

10
Dry basin of La Frescoule at Vitrolles, France (13). Example of a basin
integrated in a dense residential sector, used as a
'recreation area' and agreeable park. The small basin
appears large due to perspective
(SAUVETERRE photo)

11
Detention pond located in the zone of community activities, Paris-Nord II,
constructed by A.F.T.R.P. (Agence Foncière et Technique de la Région Paristenne)
(A.F.T.R.P. photo)

12
Detention pond at Melun-Sénart, France (77), integrated in a
residential and commercial sector, used for leisurely fishing;
a promenade on a platform supported on piles is visible in the
background
(SAUVETERRE photo)

13
Automatic cleaning of the underground tank at Sevran, France (93)
by a tipping bucket
(photo CONSEIL GÉNÉRAL 93)

14
Manual cleaning of an underground tank at Strasbourg, France (67). The employees
work in a confined atmosphere and are equipped with gas detectors
(AESN photo)

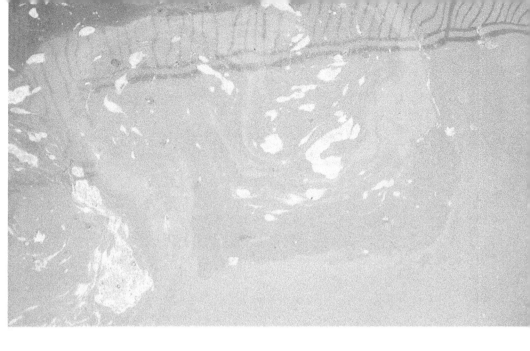

15
Debut of blooming by cyanobacteria and traces of detergents
(white patches) in a basin (SAUVETERRE photo)

16
Blooming of filamentous algae of zygnemataces
(SAUVETERRE photo)

17
Turfed 'natural' banks and a bank reinforced with
aesthetic rock garden; Melun-Sénart, France (77)

18
Turfed bank protected by rotproof piles; Torcy,
France (77)

19
Pond of Saint Bonnet, Ville Nouvelle de l'Ile-d'Abeau, France (38).
Rare example of ponds receiving stormwater from
urban areas, classified specifically as a biological
reserve. A particularly rich ecosystem
(SAUVETERRE photo)

EVACUATION OF SLUDGE

This is carried out by submerged pumps capable of carrying muddy waters with a flow rate generally between 20 and 100 l/s.

II-4.2.3 *Maintenance and management devices*

Underground retention tanks situated in dense urban zones should be of minimum nuisance and have maximum reliability of operation. This requires complex provisions for maintenance and safety and differentiates them from open basins.

PROVISIONS FOR MAINTENANCE DEVICES

Retention basins are seats of silt deposits which must be eliminated. For this purpose, the following techniques applicable to storm basins of combined sewerage systems are employed:

— manual flushing with firehoses, pressure jets etc.;
— automatic flushing by means of cisterns or tanks with valves;
— scraping the bottom of the basin. The sludge is pushed towards the outlet of the tank for evacuation. Cleaning with a scraper requires provision of an access ramp or a sloping trap for entry of machinery. Scrubbing should be done with fireproof equipment to preclude risk in case gas is present. For optimisation, this method requires:
  • access to different compartments of the tank, without disturbing flow of water,
  • adequate spacing between structural components (walls etc.) to facilitate manoeuvres,
  • external access not interfering with the usual traffic.

In all cases, the slope of the tank floor and the drainage chamber should allow easy flow of water carrying sludge:

— minimum 3% slope in the flow towards the central drain,
— 5% slope in the drain towards the sumpwell of the pumphouse.

TELESURVEILLANCE

Telesurveillance consists of transmitting to a control room all useful information concerning the state of the drainage network, machinery (including mobile equipment, pumping station, valves etc.) and eventually flow conditions (depth, flow rate). Up-to-date information about the drainage system such as defects in the position of valves, breakdown of pumps, presence of personnel etc. helps to ensure maintenance of installations, increases safety and motivates intervention by personnel. It also helps in predicting the possible occurrence of vandalism and prompts surveillance of accesses. For these reasons telesurveillance is highly recommended.

Various French organisations are equipped with some type of telesurveillance system: SIVOM* de Metz, District de Nancy, Communauté

*Syndicat Intercommunal: district

Photo 10   Flushing an underground tank using a gated valve (Saint-Dizier document).

Urbaine de Bordeau, Conseil général de Seine-Saint-Denis, City of Paris. Other towns in Europe which use telesurveillance include Berne, Breme, Hamburg, Stuttgart and Munich. Some are equipped with supplementary devices for real time control (RTC).

VENTILATION AND OTHER FACILITIES

Problems of foul odours is a big nuisance for the project supervisor. However, experience has shown that the gravity of the problem is considerably reduced with installation of efficient ventilation and quick removal of sludge after rainfall. To prevent the accumulation of nauseating or toxic gases, ventilation of underground tanks should be so planned as to allow adequate circulation of fresh air. Since experience in this domain is limited, as a precautionary measure most basins constructed nowadays are designed to be equipped with devices to treat foul odours.

Forced ventilation is also carried out sometimes before the cleaning team enters the tank; such staff are obliged to carry a device for illumination and detection of gases. This does not, however, dispense with the need for permanent sensors for explosive or toxic gases to limit the risk of intoxication or fire during visitations by maintenance personnel or in the case of accidental pollution.

MAINTENANCE AND MANAGEMENT

III-1 MAINTENANCE

Stormwater retention basins should be well maintained to ensure their durability and proper functioning. This is a basic condition for their efficient performance and acceptance by the public.

The maintenance required depends on the type of basin. Open basins, for example, especially detention ponds, constitute a lively ecological environment. Furthermore, they are often available to the public and should thus be integrated into the urban fabric; this necessitates maintenance of a certain aesthetic quality and preclusion of nuisances (foul odour, unpleasant water colour etc.).

Maintenance must be regularly carried out, which implies year-round intervention. **It should therefore be planned in detail right at the design stage of the works.**

Maintenance and the upkeep of stormwater retention basins should be planned **pragmatically**, i.e., be open to modifications based on periodic review of their state and operation. Some conditions of operation can vary so greatly that general rules with respect to frequency of intervention or quantum of sediments to be removed are not always applicable.

III-1.1 Routine Maintenance

III-1.1.1 Open basins (ponds)

It is advisable to regularly carry out
— maintenance of hydraulic works,
— maintenance of pretreatment devices,
— cleaning of basin and its environs,
— control and management of vegetation

MAINTENANCE OF HYDRAULIC WORKS
Most of the actions to be taken are dictated by common sense:
— remove flotsam and material clogging screens, weirs, orifices, scumboards etc.;

— maintain the spillways clear and free for passage of water during an unusual storm event (removal of obstructions, weeds etc.);

— replace or repair any worn-out components; ensure the upkeep of evacuation pumps;

— plan steps to prevent corrosion; check watertightness.

Valves and other devices of hydraulic regulation should be operated periodically (say once a month) to preclude jamming and blockage. Similarly, the water-level indicators in the basins should be checked to minimise the problems of drift and breakdown.

Note: Telesurveillance enables remote control and thereby optimises maintenance operations and performance of works.

MAINTENANCE OF PRETREATMENT DEVICES

Failure to properly maintain the pretreatment equipment installed on the inlet conduits and at the entry to the basins can result in total loss of performance.

Screens, grit chambers, settling tanks, oil and grease traps—all need to be regularly emptied and pollutants cleaned out.

Twice-a-year cleaning is generally a minimum for oil and grease traps, grit chambers and settling tanks. Screens should be cleaned after every heavy rainfall to obviate clogging.

CLEANLINESS OF BASIN AND SURROUNDING AREA

The users and residents of the area around a water body are most concerned about visible pollution. Maintaining an aesthetic quality in such basins is a primary objective. It requires frequent—sometimes as often as once a week, **depending on the setting of the basin surroundings**—removal of all flotsam, waste papers, bottles and other types of garbage. These materials often accumulate under the action of wind near the downwind banks and can be removed from the non-plant belt of the bank. Ancillary installations associated with utilisation of the water body (quays, fishing spots, pontoons) must also be kept accessible, clean and safe.

Dry basins should also be kept clean to:

— conserve their aesthetic quality and ancillary usages (sports acitivities, for example);

— prevent nuisance (foul odours, mosquitoes etc.). Care should be taken to prevent formation of small pans or depressions in which stagnant water can be a breeding source for mosquitoes and other insects; ponds not properly maintained can often acquire the appearance of cesspools.

In the case of basins with a cemented floor, the sludge deposited during a rain episode has to be removed. The frequency with which this operation needs to be carried out depends on the quality of inflows and the quantity of silt: it may have to be carried out after every rainfall.

To maintain open basins free of flotsam and garbage, public co-operation must be solicited. People must be educated to the fact that the environment, in particular the waterbody and approaches to ponds, have to be kept clean. It is advisable to restrict access to a limited number of spots and to place garbage bins along paths open to the public.

An inadequately maintained basin is likely to be transformed into a wild garbage dump.

CONTROL AND MANAGEMENT OF VEGETATION

The presence and development of weeds in ponds is a normal phenomenon and indispensable for providing a well functioning ecosystems. **To preclude excessive growth, both preventive and remedial measures can be adopted.**

Preventive measures are concerned with the physical and biological factors that play a role in the growth of vegetation.

— shade reduces growth of vegetation,
— reduction in quantity of nutrients available, e.g. nitrogen and phosphorus, also reduces growth.

If growth of vegetation is not controlled, insurmountable difficulties may develop, for example encroachment upon and reduction in useful capacity of the basin or subsequent plant decay which is disastrous for the water quality.

Even a rudimentary knowledge of the behaviour of aquatic plants helps in exercising reasonable control of vegetation associated with a water body. During the period of their growth a visual watch over plants is kept; species growing too rapidly or appearing in undesirable spots are manually removed. Sustained and timely action is mandatory; if undesirable elements are allowed to grow into a compact clump, uprooting them becomes difficult.

Plant thinning must be carried out every year (see III-2.2.1) in order to:

— limit progressive filling of the basin by inflow of vegetal debris,
— prevent putrescence and inflow of organic matter whose degradation consumes oxygen,
— contribute in some measure to purification of the environment by removal of biomass produced from the water nutritive elements.

When the population of microphytes (algae) becomes dense, they are collected by means of special nets or gathered at a particular spot by passing a net and removed by pumping. Special vegetation collecting boats are used for simultaneous mechanised collection of filamentous algae on the surface and in water. Removal of such algae is necessary for aesthetic reasons also.

In general, plants control should be carried out in the spirit that 'the effect of any action which is contrary to nature will cost a lot and finally will not be very effective' (SAUVETERRE [74]).

The **remedial actions** needed when dysfunctioning of retention basins is observed, are described in Section III-2.2.

Dry basins are maintained as green areas or meadows: regular shearing or mowing is essential. The products of shearing, the dead leaves (useful for compost), are collected and large branches of trees felled by wind removed. After filling, the support at the bottom of the basin may become weak; one must wait for the terrain to dry.

III-1.1.2   Underground tanks

Maintenance and management of underground tanks is more complicated than that of open basins because of:
— lesser accessibility,
— stricter constraints of safety (toxic or explosive gases),
— confinement,
— need to remove sludge deposited at the bottom of the tank.

Various auxiliary devices and equipment listed below have to be maintained in good state of operation:
— devices generating energy (power services, generators);
— valves, flaps, weirs, automatic screens;
— devices for detection of gases and associated systems of alarm: exposimeter, detectors for oxygen, H_2S, CO, CO_2 etc.;
— system of forced ventilation;
— devices for cleaning the tank; evacuation pumps;
— devices for lighting the accesses and eventually the interior of the tank.

For considerations of safety, problems of accessibility and works in the interior of underground tanks, it is beneficial to use telesurveillance and remote control to ensure:
— detection of defects and the state of various components with strict vigilance over gas detectors and devices of forced ventilation;
— testing of electromechanical equipment (valves, pumps), especially before the storm season, to ensure their reliability;
— programming an evacuation of the tank by adjusting the discharge depending on the height of water in the tank;
— keeping a watch over the rate of filling up during a storm episode; detection of hydraulic and operation anomalies etc.

Cleaning of the bottom is a crucial factor in the exploitation of an underground tank and hence the importance of a good design; experience has shown that it is very difficult to remedy the design defects of an underground tank during use.

Frequency of cleaning depends on the type of work and the manner of its feeding (the number of storm events in which it get completely filled); but as a general rule, underground tanks are cleaned after every heavy rainfall.

Primarily, it is a matter of pushing the sludge towards a storage zone, manually or by means of automatic devices, as described in Section II-4.2.3

and shown in photo 13. In the case of manual cleaning, hosepipes working at 6-7 bar pressure should be used and positioned one at every 20 metres. The network of pressurised water should be protected by a disconnecting device. Water consumption for each cleaning routine is of the order of 50 to 70 l/m² depending on the amount of deposit and the slope of the floor.

The sludge is removed from the bottom of the storage pit to the surface by an aspirator or a sludge evacuation pump.

III-1.2 Follow-up of Dyke Behaviour

Besides the usual cleaning routine described above for various types of basins, open tanks serve as an example for a follow-up of the degradation due to aging of different works. Dykes need more particular supervision. Aside from critical damage due to overtopping, the main causes of dyke failure include regressive erosion that impairs the pipes, embankment slippage etc.

The degree of complexity involved in the upkeep of a dyke depends on the size of the work and the damages likely in the case of failure. It is imperative to plan regular visits and inspections right from the day the work is commissioned, to ascertain immediately whether problems with functioning exist. In the particular case of large-size dykes involving significant risks downstream, regulations relating to large dams must be adhered to (see French circular of 14 August 1970, no. 70/15).

III-1.2.1 *Surveillance*

Essentially visual, surveillance of works constitutes a very important element of subsequent actions. It consists of a detailed annual inspection of the ensemble of works, supplemented if necessary by some tests and measurements. Qualitative inspections, done always by the same person if possible, should lead to the preparation of a report regarding the main particulars of the life and performance of the basin: date of construction, significant past incidents, repairs, filling up, evacuation etc. All parts of the basin should be minutely inspected and various anomalies noted, viz.:

— Geometrical: defects of alignment, verticality, subsidence, shrinkage, loosening, tilting etc.
— Structural: cracks and surface fractures in the work, on the causeway or in the crest of the embankments; holes and channels caused by animals, especially rodents; eventual destruction of blocks and rocks under the action of frost or physicochemical attacks.
— Hydraulic: beginning of rupture on the slopes, humid zones developing on the downstream of the embankment directly or indirectly visible (by differential growth and changes in the nature of plants), or resurgence and seepage of water downstream of the work. Watertightening membranes and joints should be checked (have they outlived their

life?). Also, actual leakage should be compared with the values obtained from computation (location, flow rate) to ascertain whether it has increased to such an extent as might endanger performance of the work. The hydraulic balance may become adverse and serious disorders may appear. If the outflowing water carries fine particles, this is an indication that the phenomenon of quicksand, consequent to piping has set in.

If the leakage becomes significantly larger than that provided for, the system of regulations should be checked first. If it is found to be in order, the increased leakage should be considered an indication of degradation of the work and complementary investigations promptly employed to determine the location and amount of leakage as well as its effect on the structure of the works. Remedial steps such as strengthening watertightness upstream, or providing additional filters and drainage downstream, should be initiated.

III-1.2.2   Instrumental survey

This survey comprises a set of measurements concerned with the behaviour of the dyke or associated works, surface displacements, internal displacements, hydraulic phenomena etc.

The plan of implementation of these measurements depends primarily on the recommendations made by the project supervisor at the design stage of the work. The measurements may also depend on the observations made during inspections. The important measures to be implemented are listed in Table III.1.

Table III.1   Measurements during instrumental survey of a dyke

Measurements	Objective
Hydraulic pore pressures	Study of stability of the body of the dyke
Compression stress	Study of vertical movement or settlement
Inclinometry	Study of horizontal movement or displacement
Clinometry	Study of rotational movement or tilting
Topography	Study of change in topography

III-1.3   Ecological Functioning of Water Bodies

The aspects discussed in the following subsections complement those given in Section I-5.2.

III-1.3.1   Recapitulation of ecology

A water body is not just an inert mass of water: it is a living medium in which large communities of flora and fauna develop. The ensemble of these living beings constitutes **the biocoenoses** whose relationship with the environments, **the biotope**, is very complex. **The ecosystem** is the ensemble formed by the biocenoses and the biotope: it is divisible into a multitude of **ecological niches**, each suited to the growth of a particular group of living species.

BIOTOPE

The biotope consists of numerous components, some exerting a dominant influence on the functioning of the ecosystem:
— **climate** controls the physical, thermal and mechanical exchanges between water and atmosphere, notably solar radiation, wind and rainfall;
— **morphology** of the waterbody: the shape, depth and slope of banks exert a direct influence on its ecological functioning. It is beneficial to maximally diversify the environment in order to promote harmonious growth of flora and fauna;
— **nature of the bottom** (edaphic factors): thickness of silt, bottom materials (rock, sand or clay) organic material deposited—these factors also determine the distribution of living species;
— **quality of water**: its chemical composition (organic matter, nutrients: nitrogen and phosphorus and micro-pollutants) determines the nature of the organisms which can subsist in it. This chemical composition depends on the volume of the retention basin and the rate of inflow: the duration for which the water stays in the basin is a determinant parameter of the physical and chemical phenomena occurring in the water body.

BIOCOENOSIS

The biocoenosis found in bodies of fresh water comprises a very large variety of organisms:

The **flora** ranges from microscopic algae to semi-aquatic plants and trees on the banks.

Algae (phytoplankton) are microscopic plants which have no roots nor means of self-locomotion. They are unicellular, grouped in colonies, in filaments which grow in the ensemble of the basin, in water, on the bottom and on immersed rocks and submerged plants. The nature of the algal population depends on the physical and chemical characteristics of the environment, each species achieving optimal growth under particular conditions, notably exposure to sun and temperature and availability of nutritive elements. Thus successions of species are often observed throughout the year.

Some typical species of freshwater algae are shown in Fig. III.1 [68].

Among the species whose proliferation can cause malfunctioning of a water body are cyanobacteria, chladophoraces and zygnemataces; a few specimens of these species are shown in Fig. III.2.

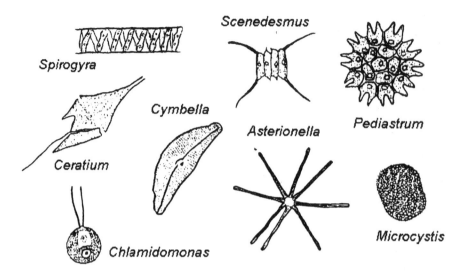

Fig. III.1   Some algae found in stormwater basins [68].

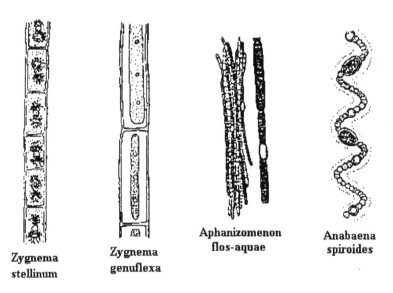

Fig. III.2   Some filamentous algae likely to give rise to algal bloom.

Aquatic plants have an organisation of individual body parts, each having its characteristic structure and function: roots, stems, leaves, flowers. Some examples were given in Section I-5.2.6.

Bushes and trees on the banks exert a significant influence on the aquatic environment: creation of shaded areas, addition of organic matter by the leaves etc.

The fauna of water basins is also rich and diversified—from zooplankton to mammals.

Zooplankton is the collective term for invertebrates which live in water; unlike phytoplankton, they are equipped with the means of self-locomotion (vibrating cilia, flagella, antenna) and can be divided into four main groups:
— protozoa (unicellular organisms, a few microns to 3 mm long),
— rotifers (size generally smaller than 0.5 mm),
— clades (shellfish 0.2 to 3 mm in size, body not segmented),
— copepods (shellfish 0.5 to 3.5 mm in size, body segmented).
Some representative specimens of the four groups are shown in Fig. III.3.

Benthic fauna, i.e., highly varied animals living at the bottom, for example sponges, hydra, worms (annelids), molluscs etc. However, arthropods constitute the most abundant group: shellfish, crayfish and insects whose numerous larvae are aquatic. This fauna is one of the primary elements of fish diet. Some representative specimens of these groups found in rivers as well as stormwater basins are illustrated in Fig. III.4.

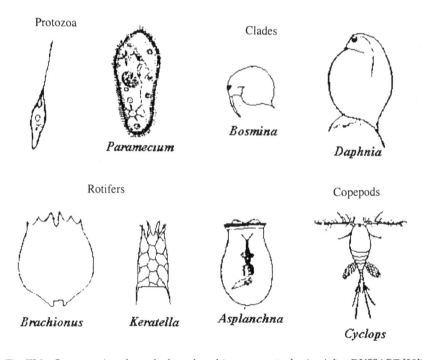

Fig. III.3 Some species of zooplankton found in stormwater basins (after DUSSART [30]).

Fishes: Of the two main varieties, cyprinoid and salmonoid, purportedly the former tolerates anaerobic conditions better than the latter.

Other animals living in water bodies or their environs which merit mention are: amphibians (tritons, frogs, toads), reptiles (freshwater tortoises,

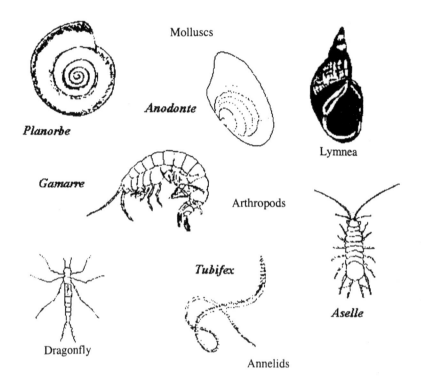

Fig. III.4   Some representative specimens of benthic fauna (after DUSSART [30]).

couleuvres), birds (ducks, herons, other waterfowl) and rodents (myocastor coypus, ondatra zibethicus, arvieola amphibius etc.).

FUNCTIONING OF THE ECOSYSTEM

To synthesise their tissues, plants make use of the elements dissolved in water: carbon, nitrogen, phosphorus, silica, magnesium, oligo elements etc. By means of their chlorophyll they are able to produce living organic matter and hence have been designated 'primary producers'. This reaction, known as **photosynthesis**, requires energy obtained basically from sunlight and is accompanied by the production of oxygen:

$$\text{Water} + CO_2 + \text{Mineral Matter} \rightarrow \text{Vegetal Matter} + \text{Oxygen}$$

The reaction is accelerated by rise in water temperature. The equation highlights the role of plants in assimilation and transformation of mineral matter; it is particularly sensitive to concentrations of nitrogen, phosphorus and silica, which show significant variations with change of season.

But plants also consume some oxygen during respiration and release corresponding amounts of carbon dioxide (CO_2), which explains the changes in pH. The processes of production of oxygen by photosyntheis and its consumption by respiration occur concomitantly, with photosynthesis ceasing, however, at night (due to absence of light). This explains why in summer, when the water is warm and therefore contains a smaller quantity of dissolved oxygen, the respiration of plants can, towards the end of night, cause an oxygen deficiency entraining fish mortality.

The variation in water temperature during a day/night cycle and accompanying change in quantum of oxygen dissolved in water is illustrated very simply in Fig. III.5.

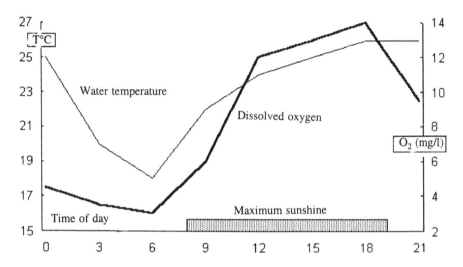

Fig. III.5   Changes in quantum of oxygen dissolved in water in the course of a summer day (after BOWMER et al., 1984).

Plants serve as food for herbivorous animals (primary consumers): zooplankton, molluscs, fish—themselves eaten by secondary and tertiary consumers: omnivorous fish, carnivores, birds, mammals etc. **The quantity of biomass produced at a particular trophic level is about one-tenth of that consumed from the lower trophic level.** For example, about 100 kg of benthic organisms are needed to produce 10 kg of fish.

Micro-organisms (bacteria, fungi) decompose and transform the organic waste into mineral products, thus completing the cycles of various elements in water. Mineralisation takes place primarily at the bottom of basins where organic matter gets deposited. It is accompanied by consumption of oxygen in the water layers close to the sediment and sometimes (for sediments rich in organic matter) also by fermentation releasing methane (CH_4), ammonia (NH_3) or hydrogen sulphide (H_2S) with its characteristic rotten egg odour.

The above ensemble of complex relations constitutes the food chain shown very schematically in Fig. III.6.

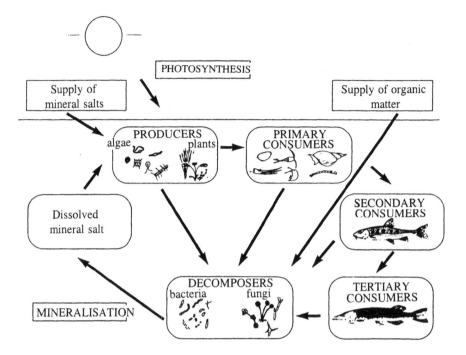

Fig. III.6   Cycle of production-destruction-mineralisation of biomass occurring in a stormwater basin (ROSENBAUM [68]).

In the water body, relationships among various species are also normally dominated by the phenomena of predation and competition: a species in competition with others does not proliferate, while one without predators will invade the water body by consuming the entire available stock of nutrition.

For a water body to function properly, a wide variety of biotopes and biocenoses should be maintained: the vegetal and animal species which mutually compete, reach an equilibrium and finally settle into the habitat they find most favourable, also called an 'ecological niche'.

In new water bodies, life is not very diversified: this is the case of a stormwater wet pond after it is filled the first time. With the passage of time, the species grow in number, compete among themselves and enrich the environment. The basin ecology evolves rapidly towards an ecological stage of regular production of vegetal and animal biomass. This evolution of eutrophic character is a natural phenomenon. In a natural pond it is slow while in a stormwater wet pond it may be very rapid due to inflow of nutrients.

Moreover, the ecological equilibrium of a stormwater wet pond is more precarious than that of a natural pond. Any change in the environment (temperature, quantum of oxygen, availability of nutrients etc.) suffices to break some links in the food chain, which disturbs the balance of the ecosystem. This is called the 'law of minimum factors', or Liebig's law. The balance of the ecosystem can also be disturbed by disorderly growth or proliferation of one of the component species.

Thus it is important to promote diversification of the biocoenosis as soon as the pond is filled the first time so that no species dominates and each species is regulated by the supply of nutrition on the one hand, and by its predators on the other. This diversification is greatly facilitated by that of habitat, especially by increasing the developed length of banks.

The common ecological dysfunctioning of water bodies is generally caused by excessive growth of the vegetal population, rarely of the animal population. The higher the buffering power of the basin, i.e., the more diversified the biocoenosis, the lower the risk of growth of any single species at the cost of others.

III-1.3.2 *Methodology of follow-up and diagnostics*

The upkeep of an ecosystem is a delicate matter due to the mutual interference among biological, chemical, climatic and edaphic phenomena. In this domain too, pragmatism is the rule. As in the case of any artificial environment which should evolve towards a eutrophic stage in nature, only ecological upkeep of ponds through periodic observations enables timely detection of disorders, and even anticipation of their occurrence and implementation of simple remedial actions.

Routine surveillance should comprise inspection of the following:
— water surface for algal bloom, undesirable colour or iridescence, filamentous streaks, floating bodies etc.;
— behaviour of animals, fish, birds etc.; it may be useful to consult the residents, especially fishermen, who are often close observers of changes in the basin;
— turbidity of water: a simple measurement by "SECCHI" disc gives a good indication of planktonic changes;
— measurement, at first in-situ, of some simple chemical and biological indicators such as dissolved oxygen, pH, conductivity etc., followed by laboratory analysis, if necessary;
— identification of vegetal species present, occupied zones, their rate of growth (a count at regular intervals).

All observations should be recorded in a diary or in the pond-dossier (see Introduction), which will:
— facilitate its management,
— be useful as a reference during the design of other basins in climatically or morphologically similar zones or for correction of faults observed in other installations.

For basins of simple shape and limited size, the analysis can be carried out on a single sample taken at the outlet, at a depth of about 1 metre below the water surface; it will be reasonably representative of the water leaving the basin and for most of the parameters of the large mass of water in the basin. When the basin size is larger than ten hectares, it is preferable to take samples at several points.

To take into account variations due to the annual biological cycle, it is desirable that observations be made routinely thrice a year—in spring, at the end of July and in September.

The nature and importance of the follow-up operations of the ecosystem depend certainly on the usage of the basin; the more demanding the needs, the larger the number of such operations.

Detailed ecological examination of a water body is undertaken when there are operational problems. This task has to be entrusted to specialists (ecologists, chemists, biologists). Tables III.2 and III.3 summarise the minimal particulars which must be noted respectively in the:

— basin data sheet,
— preliminary report of the diagnostic examination.

Table III.2   Typical data sheet for a stormwater detention pond

DATA SHEET FOR WATER BODY	
Identity	
Name of basin:	
District:	Community:
Proprietor:	Date of commissioning:
Manager:	Type of basin:
Morphology	
Surface area:	Type of outlet:
Volume:	Flow-control devices:
Perimeter:	Dyke details:
Drawdown range (frequency):	Cross-flow (if any):
Maximum depth:	Position with respect to water table:
Average depth:	Types of banks:
Primary functions	**Ancillary Usages**
Food management:	Beautification of area:
Remediation of run-off:	Fishing:
Others:	Sports:
	Others:
Catchment Area	**Treatment systems**
Total surface area:	Pretreatment:
% residential:	Post-treatment:
% pavilion:	Oxygenation:
% commercial:	Compartmentalisation:
% industrial:	Watertightness:
% infrastructures:	Others:
% rural:	
% forest:	

Table III.3 Example of a diagnostic report with normal range of values of various parameters and symptoms in case of anomaly

ECOLOGICAL DIAGNOSIS OF WATER BODY

Name of basin
District   Community
Date of visit   Time
Meteorological conditions
—on that particular day
—during the preceding 10 days
—location(s) from which sample(s) taken

	Parameters	Units	Normal range of values	Remarks
Water	transparency	cm	30 to 100	When < 30, decreased photosynthesis
	temperature	°C	5 to 25	
	DO	mg/l	6 to 13	When < 1.5, lethal for cyprinids
	DO-saturation	%	70 to 150	When < 30-40, asphyxiation
	pH		6.5 to 9	
	conductivity	µS/cm	400 to 600	
	SS	mg/l	15 to 40	If high, causes silting and turbidity
	Oxydability	mg/l	4 to 15	Indicates consumption of oxygen; when high, warns of fish asphyxiation
	BOD_5	mg/l	2 to 10	If high, indicates inflow of waste water
	COD	mg/l	10 to 30	Same
	N_{total}	mg/l	1 to 10	Plant infestation if > 15
	$N-NO_3$	mg/l	1 to 5	Indicates pollution, generally by sewage
	$N-NH_4$	mg/l	0 to 1	Indicates pollution, generally by sewage
	Chlorides	mg/l	20 to 65	Causes death of phytoplankton if high
	P_{total}	mg/l	0.05 to 1	
	Anionic detergents	µg/l	0 to 50	If > 50, inhibits self-purification
Silt	Organic matter	%	2 to 15	High organic matter indicates insufficient mineralisation
	P_{total}	mg/g	0.2 to 4.0	If high, release possible
	Lead	mg/kg	30 to 100	French standard 44–041 < 300
Plankton	Species	Number	> 15	If number < 5 to 10, risk of pullulation of a species

If plankton too abundant, turbidity will increase

Plants	Area covered by hydrophytes	% covered		Increase in organic matter may occur
	Species	Number	diversity	
Benthos	Inventory	Number	diversity	This observation gives a good idea of potential for fish development and biological quality of the environment

Criteria of evaluation: if only the water quality is considered, one may refer to the 'description of the water quality and watercourses' as defined in the French circular of November 1971. Five categories of water are defined on the basis of physicochemical and biological parameters (1A: very good, 1B: good, 2: fair, 3: bad and HC: very bad).

III-1.3.3 Management of fish populations

Wet ponds eventually become naturally stocked with fish, especially if crossed by a stream. However, if recreational usages (beautification, fishing) are to be provided, the basin should be stocked with fish one year after commissioning, preferably by introducing fingerlings during spring. Stocking directly with adult fish (ex.: for fishing competitions) is not a good idea as a high mortality rate may ensue due to difficulties in adaptation.

For a stable fish population, introduction of the following fishes is recommended:

— a good phytophage, such as the roach[1], the gardon[2] (strictly speaking, the latter is not phytophagous);
— carnivores such as the pike[3], sandre[4] perch[5] (remembering that the latter is very prolific);
— a detritivore, say tench.[6]
   'Non-utilitarian' (carassius[7], rainbow perch[8]) or harmful (catfish)[9] fishes should be avoided. Another fish to eschew is the carp[10] because it scrapes silt from the bottom and thereby increases water turbidity; this in turn impairs the growth of aquatic plants and, furthermore, spoils the aesthetics of the basin.

Management of the fish population involves:

— Study of species distribution. For various reasons, some species may have difficulty in reproducing while others tend to be prolific. Such a study is readily carried out by talking to fishermen or taking an inventory during wet pond evacuation.
— Taking out of the pond, fishes by angling, electrical means or through winter fishing, with emptying the pond.

Emptying of the pond is a good time for controlling the fish population and eliminating undesirables, the 'monsters' (large pikes and carps). The population may be readjusted and, if necessary, some healthy fish removed and introduced in another pond; these operations are generally carried out in winter.

Fishing on the banks of a pond is not only a very popular activity among the locals, but is also helpful in maintaining an ecological equilibrium in the detention pond. However, one should consider not only the interests of the fishermen, but also pay due attention to the rules that ensure minimisation of eutrophication in the basin, especially by giving up bait with the result

---

[1] Rotengle, Scardinios, [2] Rutilus, [3] Brochet, Esox lucius, [4] *Sander lucioperca* [5] Perca fluviatilis
[6] Tinca, [7] *Carassius*, [8] *Lepomis gibbosus* [9] *Ictalurus melas* [10] *Cyprinus carpio*

that a large quantity of organic matter—the quantity increasing with the number of fishermen—is added to the aquatic medium.

As the initially planned objectives of the basin must continue to be fulfilled, it is recommended that each basin be placed in the charge of persons responsible for sewerage and having good relationship with various users.

III-2 MALFUNCTIONING OF WET PONDS AND REMEDIAL MEASURES

III-2.1 Common Dysfunctioning

III-2.1.1 *Pollution by waste water*

Waste water causes an accelerated and harmful eutrophication; it can transform the pond into a cesspool in a short period of time. Pollution generally results from faulty connections and consequent flow of waste water into a stormwater drain; such defective connections should be quickly traced and remedied. Waste water can also be introduced from old combined drainage networks which, in spite of due precautions, still end up in the storm-drains or directly in the pond; this situation must likewise be quickly corrected to avoid damage.

III-2.1.2 *Plant proliferation*

Uncontrolled growth of aquatic plants can generate a number of disorders in the water body.

EXAMPLE OF PROLIFERATION

Jussieua grandiflora was mistakenly introduced in some water bodies and proliferated rapidly: after two seasons, two-thirds of the surface of these water bodies was completely covered by this plant. Its stems float and quickly grow to reach the water surface where they then become aerial and form an almost uncontrollable large vegetal mass. When weeded, each piece of stem left in the water generates roots and implants on the first bank encountered. The large organic mass, associated with sedimentation that spreads over the entire invaded area, requires quick and heavy machinery (mechanical shovel) for eradication. Subsequently, any new vegetal growth must be controlled with herbicides, even though such products should, in principle, be prohibited and the water bodies preferably allowed to develop their own ecological control. But of the two evils, the lesser has to be chosen. True, the best precaution is not to plant such species in the first instance!

Basin components in the form of channels or other watery areas, narrow and often shallow, are very susceptible to invasion by higher aquatic plants. Special attention must be paid to the genera *Phragmites*, *Typha* and *Scirpus*. They release rhizomes—floating type in *Phragmites* and bottom creepers in

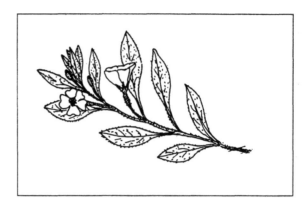

the other two genera; these rhizomes produce new stems and the chain continues until the entire medium is invaded. The channel fills up quickly because the roots entangle and new ones grow on the organic waste of the old. Such disorders are almost unavoidable if the canal is shallow. Even otherwise, the floating rhizomes of *Phragmites* may cover a water breadth of 8 to 10 metres. Further, some species (*Scirpus lacustris, Typha angustifolia*) may grow in water two or more metres deep. The last two species must be watched closely and permitted to develop only in scattered islets. The remedy is easy if timely applied: the new sprout released by rhizomes, and that isolated in the water, should be pulled out.

The species *Elodea canadensis* is photosynthetically very active and participates in oxygenation of water; in moderate quantities its action is beneficial. Earlier it tended to pullulate but during the last fifteen years this tendency has diminished. Today its presence poses no serious problems; it has receded or even disappeared from some basins.

Duckweeds* are another example of invasive plants. They can fully cover a waterbody and form a several centimetre thick bed that blocks light from the lower water layers, causes a rapid drop in oxygenation and engenders the phenomenon of anaerobicity in the benthic silt: transformation of nitrates to nitrites and release of ammonia and even hydrogen sulphide (with rotten egg odour). Under such conditions, fish may die of asphyxiation. Duckweeds are more likely to grow in small ponds and canals sheltered from winds, rich in organic matter or nitrates, under dense covering of shrubs or under forest. These weeds are always difficult to eradicate.

WATER BLOOMS

Blooms of filamentous algae form a large mat that can cover the entire surface of a new water body. They generally appear after a sunny period and disappear spontaneously a few days later through decomposition and settlement of organic mass to the bottom (see colour photo 16).

---

*lemna minor

Sometimes, an intense pollution destroying the biocoenosis or an excessive input of nutrients may induce a massive growth of filamentous algae. These phenomena indicate a shift in equilibrium which is generally temporary in the aquatic ecosystem.

Blooms of cyanobacteria form a thin layer which, under the effect of wind, resembles a coat of bluish-green paint (see colour photo 15).

Among the algae yielding blooms, mention may be made of zygnemataces (*Spirogyra, Mougeotia*) and cyanobacteria such as *Aphanizomenon, Anabaena*; exceptional blooms of *Hydrodyction reticulatum* have also been observed.

III-2.1.3   Fish mortality

Among the causes of fish mortality, let us mention the following:
— hypereutrophication of the water body, which is accompanied by a deficiency of dissolved oxygen towards the end of night;
— asphyxiation caused by excessive organic matter, especially during the storm period (the well-known shock effect in rivers due to inflow of storm waste waters via combined sewers).
— inflow of toxic material transported by the drainage network;
— overpopulation of fish due to lack of predators;
— parasitic diseases.

III-2.2   Remedial Measures

The common dysfunction of waterbodies are generally caused by an unduly large increment in the plant population and rarely by an overabundant animal population. The primary remedy is to export the excess vegetal biomass produced.

III-2.2.1   Remedial weeding

Weeding has two objectives:
— routine maintenance of the water body by removal of surplus organic matter,
— remedial intervention to preclude invasion of the water surface by aquatic plants.

The best time for carrying out this operation is July-August but care should be taken not to disturb the niches. If carried out too early, the entire growth is not removed and the operation must be repeated. Contrarily, if carried out much later, the aquatic plants have already stored what they need in organs buried in silt (rhizomes, roots) and have become so strong that weeding is ineffectual and the intensity of growth the following spring not constrained. Late weeding removes a small portion of lignin and cellulose only, with no beneficial effect accruing to the ecosystem.

In weeding a water body, all the products must be removed to obviate their decomposition and mineralisation. This decomposition consumes part of the oxygen in the water, concurrent with the other users; it also promotes silting of the basin and restitutes the eutrophicating nutritive mineral elements in the aquatic environment. The weed products removed can be used as a soil conditioner after proper composting.

Remedial weeding does not dispense with the need for regular control of plant development. If growth is left unchecked, removal becomes a more difficult and expensive operation.

There are three methods of weeding:

MECHANICAL WEEDING

The most effective means is the weed-cutter boat fitted with a rake. In addition to devices for weeding and pick-up, this type of boat can be equipped with:

— a harrow to detach creepers and plants from the bottom,
— a rotating detriter to detach grass, rhizomes and roots etc. and partially desilt by creating a flushing flow. Cuts are simultaneously callected; this reduces loss of plant pieces in the water body and subsequent multiplication by the cuttings.

For large water bodies, subaquatic reapers mounted on boats with flat bottom and paddle wheels are best. Flotsam and chopped plants are collected in skimming baskets.

Photo 11   Weed-cutter boat (from the document *Européenne de Dragage*)

CHEMICAL WEEDING

Chemical treatment requires excellent knowledge not only of aquatic flora, but of the chemicals to be used in its control since their action is often target-specific.

The most commonly used algicide is copper sulphate, although its secondary effects are controversial. The maximum dose is of the order of 0.5 mg $CuSO_4/l$ once a fortnight during summer. Its efficacy seems to improve in slightly acidic water but as the Cu ions are toxic, its concentration in sediments should be checked regularly. It may be recalled that the best means of limiting the proliferation of algae is to prevent the inflow of nitrates and phosphates into the basin.

For higher plants—hydrophytes and helophytes—use of herbicides and pesticides in an aquatic medium should be resorted to only as a last recourse because the degradation caused by these substances to the eco-system as a whole may outweigh the benefits derived. Although some herbicides do not act directly on the fauna when applied in prescribed doses, they can influence growth and composition of plankton as well as the concentration of dissolved oxygen and the pH. It may also be noted that contrary to the case of soil in which pesticidal diffusion is slow and limited, in water diffusion is very rapid.

In France, the herbicides used in an aquatic medium are subject to official certification and the sales permit specifies the purpose for which they are to be used, viz.:

— destruction of algae,
— destruction of semi-aquatic plants,
— destruction of aquatic plants.

Table III.4, excerpted from the French 'phytosanitary index' ACTA,* lists the major authorised synthetic organic herbicides/weedicides. This index is updated every year and should be consulted** for determining the dose per hectare, content of active ingredient etc.

BIOLOGICAL CONTROL OF PLANTS

Control of plants may be (partially) effected by animals which consume vegetal biomass for nourishment. The rate of withdrawal is generally small and the objective can only be a support to control of vegetal growth. Phytophagous fish and palmipeds participate in this natural process.

III-2.2.2 Aeration

The objectives of aeration are to maintain a mimimum concentration of oxygen to combat nuisance-causing anaerobic fermentation and to assist in the survival of fish during difficult times, especially in summer. The various means employed are:

---

*Association of Agricultural Technical Coordination
**hopefully, there are similar documents in other countries

- release of microbubbles of air or oxygen near the bottom,
- fountains and jets of water,
- recirculation of water with cascades.

The last two techniques combine aeration and aesthetics.

One should not depend on these devices for treating the effects of a chronic pollution; at most, they tackle only localised and short-term problems of oxygen deficiency.

Table III.4 Excerpt from the phytosanitary index ACTA (Association of Agricultural Technical Co-ordination of France).

Brand	Supplier	Active ingredient	Toxicity	Susceptible plants
Weedazol TL aqua Herbaquat Torpi-aqua	CFPI RHODIC ASEPTAN	Aminotriazole	low for fauna	Helophytes
HTG 10 granu aqua	CFPI	Chlortiamide	very strong	Hydrophytes
Stymelol Alatex A	STYMELOL PEPRO	Dalapon	low for fauna	Helophytes
Aquaprop Oxyjonc (granulated) DXA 707	QUINOLEINE ASEPTAN NGA	Dichlobenil	toxic for fish if > maximum dose	Hydrophytes
Reglex 2	SOPRA	Diquat	toxic for fauna waterlilies resistant	Dicotyledonous Hydrophytes
Sonar P5	DOW ELANCO	Fluridone	low	Hydrophytes Helophytes
Roundup 360	MONSANTO	Glyphosate	low	Helophytes

III-2.2.3   Periodic emptying

Emptying of wet ponds helps:
- in carrying out maintenance of works that remain submerged in normal times,
- eventually in larger cleaning operations (rather rare).

It further enables:
- complete (or almost complete) renewal of water;
- exposure of the bottom silt to the atmosphere, thereby stimulating oxidation, decomposition and mineralisation;
- control of the fish population. To achieve this the manager requires the competent assistance of specialists (In France: CEMAGREF,* CSP,** fishing societies etc.).

---

*Centre d'Etude du Machinisme Agricole du Génie Rural et des Eaux et Forêt (one of the research centres of the Ministry of Agriculture)
**Conseil Superieur de la Pêche (council of fishing)

A time gap of 10 years between emptying operations is considered normal. Emptying operation can be planned when some other job is to be executed: repair of banks for example to rectify the slow degradation of the water quality caused by reduction in transparency or algal invasion. It may be recalled that fishing ponds are emptied, on average, every two or three years.

Emptying is preferably carried out in winter because risk for the environment is minimal in this season and fish can be manipulated without much harm as their scales are protected by a thick mucus at this time.

Aspiration dredger for shredding plants

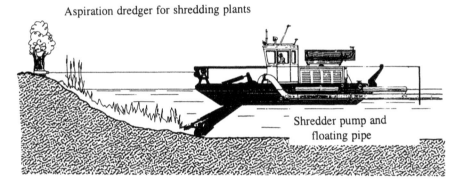

Shredder pump and
floating pipe

Fig. III.7   Aspiration dredger for shredding plants (from the document *Européenne de Dragage*).

Refilling of wet ponds is usually carried out before spring. It is imperative that the water bodies be filled before commencement of the spring biological activities.

III-2.2.4   Elimination of silt and other wastes

The occasion of basin emptying is used for clearing accumulations in front of the collectors (inlet channels) or removing waste materials deposited under the action of dominant winds along the downwind banks.

The larger operation of generalised or almost generalised cleaning of the entire bottom of a water body is very time-consuming and except for accidents, is carried out rarely, because silting is usually a very slow operation. Accidents may arise from:

— rapid silting-up from the catchment area when works are in progress. When the catchment area contains a large quantity of loose earth, a violent storm can blow much of it into the basin. Precautions have to be taken to limit the flow of stormwater during construction of large works (for example, straw packs/bags can be heaped here and there to filter the incoming water) or efficient settlers installed on the major brooks or closed collectors.

Photo 12   Cleaner boat

— untimely reed growth. Reeds thrive on their wastes, the roots entangle and clogging of the pond can occur quite rapidly—at the rate of several centimetres, even a decimetre per year.

MECHANICAL DEVICES FOR DESLUDGING

Sludge-removal operations are carried out under dry conditions after emptying or under water with the help of a suction device. In the absence of settling areas close to the pond for the drying operation (usually odoriferous), the sludge should be hauled away quickly when so required by the environment: inhabited or commercial zones etc.

PHYSICOCHEMICAL CONTROL OF SLUDGE

Physicochemical procedures can reduce the quantity of silt without emptying of the pond: chalk, lime or freeze-dried bacteria can be spread on the surface of water. But these techniques are applicable only in the case of highly organic silts whose degradation is desired to be accelerated. Eventually these techniques should permit spacing out desludging operations. But they presently do not guarantee positive results and applications must be regularly repeated to obtain visible effects. Research on the efficacy of such procedures is underway.

III-3 DISPOSAL OF SLUDGE AND WASTES

Evacuation and disposal of various wastes and sludge are serious problems. They should be considered right at the design stage of the basin, keeping in view the regulatory requirements.

In practice, there are very few solutions available today; a dump or waste water treatment works are still the main alternatives. Nevertheless, the problem is under serious study. For example, with respect to sludge, the techniques of separation, desiccation and recycling are being tested.

Wastes collected in pretreatment works should be transported to a dump or treated in a special treatment unit (especially hydrocarbons).

Floating and residual matter of various kinds collected during cleaning of ponds and their accesses, as well as the plant material obtained from shearing or weeding, should be transported to a dump; the latter, of course, could alternatively be composted.

The major problem, without doubt, is the sludge that accumulates at the bottom of the basin; several solutions are available:

— Evacuation by gravity or pumping of liquid sludge into the sewerage system. Intermittent and massive inflow of this sludge may disturb the functioning of the waste water-treatment unit. Therefore the feasibility of this solution should be checked before implementing it. Transport by the sewerage network can be planned only if the hydraulic conditions are favourable (velocity higher than 0.3 m/s for preventing deposition in sewers).

— Transport by special trucks to an installation of dehydration associated with a waste water-treatment works. As a matter of fact, it is often not economical to transport the already concentrated sludge through a sewer for complete retreatment in the waste water-treatment unit; truck transport may be cheaper.

— Transport the sludge collected through a cleaning operation to a central treatment unit (case of Seine-Saint Denis in France) where it is dehydrated before transference to dumps.

— Disposal on agricultural land, alone or mixed with sewage sludge.

Before planning agricultural recycling of pond sludge mixed with sludge from the waste water treatment unit for example, it is advisable to carry out physicochemical analysis as proposed by AFNOR NF U44-041 (French Standard), which prescribes the characteristics and possibilities of sludge use depending on its composition (especially heavy metals). Table III.5 summarises various current possibilities for waste disposal.

Table III.5 Destination of wastes as a function of their origin

Waste	DESTINATION			
	Dump	Waste water treatment works	Specialised treatment unit	Others
Residues from pretreatment devices				
Sand	X	X		
Collected waste matter	X	X		
Hydrocarbons			X	
Residues from cleaning basins and approaches				
Flotsam	X			
Surplus plant material	X			Compost
Sludge deposits	X	X	X	Agricultural recycling
Silt	X			Agricultural recycling

III-4 ACCIDENTAL POLLUTION

The contents of this section are based on the results of an interagency* study entitled 'Accidental Pollution of Inland Waters' (1991) dealing with serious accidents. It can inspire follow-up measures for the pollution frequently encountered in stormwater detention ponds.

Stormwater networks are often highly susceptible to accidental pollution. Such pollution may be caused even by vehicular traffic as well as by various other activities carried out on the catchment area.

Accidental pollutions may be of different types. The most common are those caused by spillage of hydrocarbons or phytosanitary products (pesticides) and metallic pollutions. They constitute grave risk for the ponds as well as the downstream receiving waters. Other pollutions, such as sludge from construction works, essentially of the mineral type, may be relatively less harmful. The risks involved should be taken into account right at the design stage, by planning the accesses, works and equipment in such a manner as to facilitate intervention.

When an accident occurs, appropriate precautions should be taken on the one hand for human safety and on the other, for containing the extent of pollution; in short, one should always be prepared for the eventuality of such situations.

III-4.1 Combat Preparation

Departmental procedures for intervention and other exigencies have been

*Water Agencies (whose territory extends to the water catchment area of the main rivers: Seine, Rhône, Rhin etc.) forming part of the Public Management of Water Resources in France.

prescribed by the prefectorial services in France (Decree no. 88-622 dated 6 May 1988 concerning emergency procedures in accordance with the law 'Civil Safety' no. 87-585 of 22 July 1987).

Fortunately not all accidental pollutions are of such intensity as to warrant activation of stipulated emergency procedures. Nevertheless, an appropriate plan of operation should always be available to the local (community or intercommunity level) bodies.

To determine the crucial operational procedures that can be quickly and effectively applied, the following information is required:
— the routes on which pollutants are to be transported,
— installations of industrial sources of pollutants,
— list of water tappings and pumping stations,
— map of vulnerable points of groundwater tables,
— list of possible accidents and those that have occurred in the past,
— plan of sewerage system,
— map of the transit time of pollutants in the sewerage network,
— list of recognised laboratories for analysis of water etc.

Keeping in view the existing organisations and facilities (personnel, materials), a study of the scenario will facilitate planning the appropriate procedures and assignation of responsibilities.

III-4.2   Observations

The quality and accuracy of the early observations are determinant factors for the speed and efficiency of the measures to be undertaken.

WHAT SHOULD BE THE FOCUS OF OBSERVATIONS?

UNUSUAL GENERAL FEATURES:
   surficial, in the water mass, at the bottom of the water body
COLOUR
ODOUR
PARTICULAR MANIFESTATIONS
• dead fish or animals
• dead plants
• foam, froth
EXTENT
DEVELOPMENT
PROBABLE ORIGIN

III-4.3   Sounding an Alert

A flow chart for sounding an alert in case of a pollution episode is given in Fig. III.8.

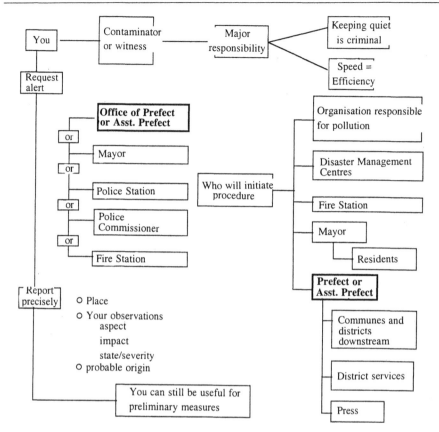

Note: Residents around a basin are privileged observers. Keeping them well informed about the action to be taken in case of accidental pollution can be an effective means of sounding an alert.

Fig. III.8 General flow chart for an emergency alert in case of accidental pollution.

III-4.4 Evaluation

In cases of accidental water pollution, one must closely examine the situation as it develops, plan follow-up action and gather all possible relevant information. Several parameters influence propagation of pollution:

— physicochemical behaviour of the pollutant in water (Fig. III.9),
— wind,
— hydrodynamic parameters such as turbulence.

III-4.4.1 Identification of pollutant

Identification of the pollutant can be carried out directly if the origin or source is known or the pollutant readily identifiable. When such is not the case, an analytical identification becomes necessary. The representativeness of the samples taken for analysis is of fundamental importance.

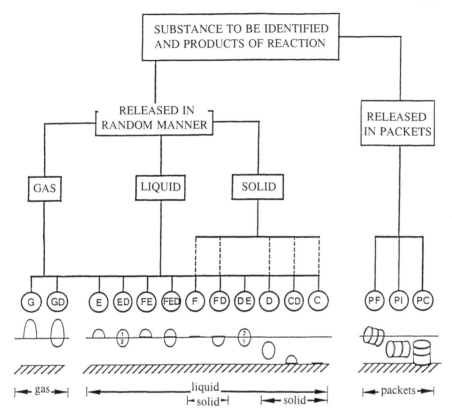

Fig. III.9 Normalised European system of classification of pollutants (Bonn Accord: BAWG-OTSOPA).
C: flowing product, D: soluble product, E: volatile product, F: floating product, G: gas, I: between two water containers, P: packets

III-4.4.2 Sampling

Proper choice of the number and location of points from which samples are taken should enable correct identification and estimation of spread of the pollution in space and time.

In many cases taking samples of the pollutant or the polluted water from the basin as well as the network, constitutes the first step for the responsible authorities in evaluation of the situation and development of a strategy to combat it. If fish mortalities have occurred, dead fish constitute critical samples.

The samples should be transported to the laboratory as quickly as possible. Recommendations for preservation of samples are given in various documents and guides and should be meticulously followed.

III-4.5   Intervention

III-4.5.1   *Public safety and workers' safety*

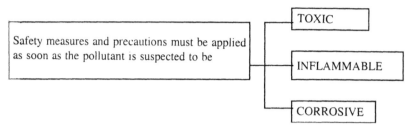

Safety measures and precautions must be applied as soon as the pollutant is suspected to be

TOXIC

INFLAMMABLE

CORROSIVE

III-4.5.2   *Measures for safeguarding the basin*

Safeguards include the following:
— measures aimed at limiting the impact, for example:
   * lower or raise the water level in a pond before taking out the pollutants to be collected,
   * divert the pollutants towards a less sensitive zone or another drainage network,
   * reduce the time of transit through the basin, taking care not to adversely affect the downstream areas;
— operations of 'rescue' of the ecosystem, such as oxygenation of the medium, electrical fishing, organisation of displacement of living communities etc.

III-4.5.3   *Measures to combat episodal pollution*

Easy access to basins helps in implementing counteractive measures. Disaster relief will be more efficacious if some preventive steps have been taken in advance: anchorage for floating barrages, dead bodies, stationary installations of combat or protection (floating barrages, pumping systems, confining sheets).

MEASURES FOR CONFINEMENT OF POLLUTION TO LAND
In view of the fact that most human activities take place on land, only exceptionally would an accidental pollution occur directly in a water body. Therefore, interventions on land are more effective than those carried out on or in water. Remaining within the stipulated rules of safety, every effort should be made to take advantage of the time and space between the place of accidental release of the pollutant and the water body. The measures to be implemented on land are explained in Figs. III.10 to III.12 and their objectives are:
— to dry up the source of pollution,
— to prevent or restrain its spread in the aqueous medium.

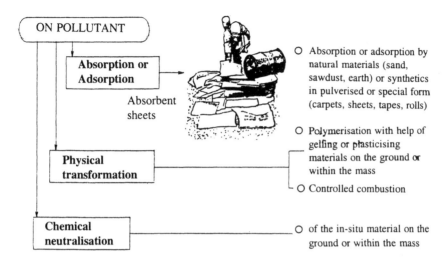

ON POLLUTANT

Absorption or Adsorption

Absorbent sheets

O Absorption or adsorption by natural materials (sand, sawdust, earth) or synthetics in pulverised or special form (carpets, sheets, tapes, rolls)

Physical transformation

O Polymerisation with help of gelfing or plasticising materials on the ground or within the mass

O Controlled combustion

Chemical neutralisation

O of the in-situ material on the ground or within the mass

Fig. III.10   Action on the pollutant

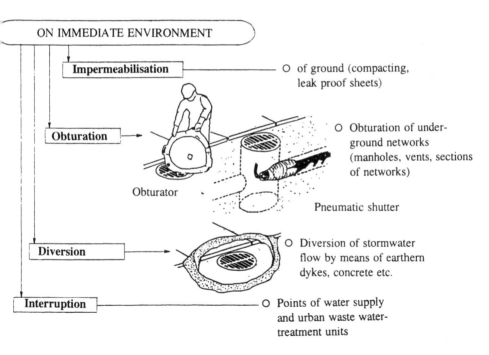

ON IMMEDIATE ENVIRONMENT

Impermeabilisation

O of ground (compacting, leak proof sheets)

Obturation

Obturator

O Obturation of underground networks (manholes, vents, sections of networks)

Pneumatic shutter

Diversion

O Diversion of stormwater flow by means of earthern dykes, concrete etc.

Interruption

O Points of water supply and urban waste water-treatment units

Fig. III.11   Action on immediate environment.

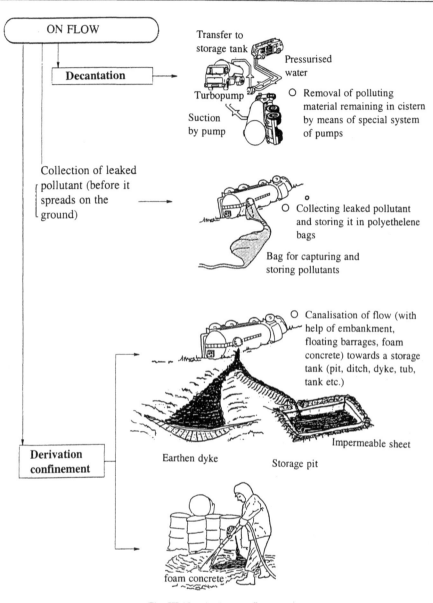

ON FLOW

Decantation

Transfer to
storage tank

Pressurised
water

Turbopump

Suction
by pump

O Removal of polluting
material remaining in cistern
by means of special system
of pumps

Collection of leaked
pollutant (before it
spreads on the
ground)

O Collecting leaked pollutant
and storing it in polyethelene
bags

Bag for capturing and
storing pollutants

O Canalisation of flow (with
help of embankment,
floating barrages, foam
concrete) towards a storage
tank (pit, ditch, dyke, tub,
tank etc.)

**Derivation
confinement**

Earthen dyke

Storage pit

Impermeable sheet

foam concrete

Fig. III.12   Action on flow.

Such measures are implemented in the spirit of the following principle:

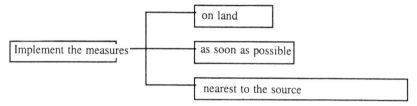

REMEDIATION, CLEANING, RESTORATION

To prevent damage to humans and the environment by the diffused residues of polluting chemicals, the operations of remediation and cleaning must be activated as soon as possible. Depending on the nature of the pollutant and the affected locales (bank, bottom of the basin, works etc.), these operations can be of various types: removal of polluted soil and plants, cleaning of works, remediation of earth etc.

TRANSFER, TRANSPORT AND TREATMENT OF POLLUTANTS

The pollutants and the polluted wastes must be transferred to a special treatment unit. Care should be taken that operations of loading and transport do not contribute to spread of the pollution; stipulations contained in the Regulations and Rules of Transport of Dangerous Substances must be adhered to.

COSTS

IV-1   INTRODUCTION

The cost of an installation, whatever it may be, consists of the following components, which must be included in the financial planning:
— preliminary studies,
— equipment and construction (capital investment),
— management and maintenance (operational or overhead costs).
The first component, 'preliminary studies', is essential if these goals are desired:
— certitude that the kind of work chosen is the best in the given situation and is not dependant on options that do not satisfactorily meet the requirements;
— maximum economising on time as well as cost during construction and planning.
It is during the preliminary studies that one can take advantage of the multipurpose utility of the installation in terms of space as well as time.

The second component, 'capital investment', comprises acquisition of land, works, equipment as well as the supervisory staff. The formalities include a preproject summary, detailed preproject study, various documents pertaining to the contract, routine and follow-up works etc.

With respect to equipment, its cost depends on the prevailing market rates. As for the supervisory staff, the rates of remuneration applicable in France are codified in CCM No. 2001-1 in the regulations text pertaining to engineering and architectural works, under the heading 'Infrastructures'.

The third component, 'operational (overhead) costs', includes the cost of maintenance as well as replacement of various equipment when their practical life is over.

IV-2   INVESTMENT NECESSARY: PRELIMINARY STUDIES

The preliminary studies were reviewed in Section I-1.2 and comprise:
— hydraulic investigation and hydrological study,

— investigation into potential sites and subsequently the sites chosen,
— impact study.

The cost of preliminary studies comprises a fixed sum for general analysis and collection of secondary data and a fluid sum which varies directly with the size of the reconnaissance tasks. Thus the cost of preliminary studies varies from 4 to 30% of the cost of the works; it was found to be close to 11% for the 21 basins studied, most of them situated in the region Ile-de-France [76].

IV-3   COST OF CONSTRUCTION

The cost of construction of stormwater basins can be split into two components: one under the heading 'engineering' and the other under 'works'.

IV-3.1   Cost of Engineering

This includes the cost of the general supervisory staff with or without the project. Table IV.1, compiled on the basis of the French 'Market Code', gives some idea of the rates applicable (percentage of the total cost) for a 'Mission m 1' of the general project management, corresponding to some values of the total cost and for varying degrees of complexity (ranging from 4 to 7%, as defined in the French Market Code CCM No. 2001-1).

Table IV.1   Order of costs of general supervisory staff of type m 1, expressed in percentage of total cost for four common degrees of complexity (from CCM No. 2001-1)

Total cost (Francs)	Degree of complexity (under the heading sewerage)			
	4	5	6	7
200,000	12.44	13.47	14.64	15.97
500,000	11.47	12.81	14.31	15.97
1,000,000	10.00	11.17	12.48	13.93
	8.05	8.99	10.04	11.21
	7.58	8.47	9.46	10.56
	6.96	7.78	8.69	9.70
	6.82	7.62	8.51	9.50
500,000,000	6.62	7.40	8.26	9.23

Mission m 1 is a technical mission, comprising all stages of design and follow-up action from 'Preproject Summary' to "Dossier (completion report) of Executed Work'. The total cost, final or provisional, is the estimated price of investment, including the cost of the engineering project management (Article 4.I of the decree 73-207 of 28 February 1973).

IV-3.2 Cost of Investment

A distinction must be made between ponds (open basins) and underground tanks because various components of the overall cost carry different weights accordingly. Also, the cost of acquisition of land is not taken into account here as it varies greatly with the situation.

IV-3.2.1 Open basins

The cost of construction can be split under the following headings:
1) Preparatory works (topography, establishment)
2) Earthwork (clearing, embankment)
3) Impermeabilisation (of bottom, of walls) and/or drainage
4) Outflow works (of evacuation, of regulation)
5) Other works (pretreatment etc.)
6) Protection of banks (pitching, antilapping)
7) Landscaping of embankment (plants, banks)
8) Miscellaneous works (equipment, paths, safety)
9) Dykes (if relevant)

To obtain some idea of the costs of these components, each was calculated as a percentage of the total cost for the 21 stormwater basins studied [76]; the details are presented in Table IV.2.

Table IV.2 Comparative costs of various component works of stormwater basins shown as percentage of total investment (excluding cost of preliminary studies and supervisory staff); sample size 21 basins [76]

Type of basin	Dry basins (7)			Wet ponds (14)		
	Max.	Mode	Min.	Max.	Mode	Min.
Preparatory works	17	3	1	8	4	0
Earthwork	63	30	15	83	60	13
Impermeabilisation	45	17	0	16	10	0
Outlet works	37	5	2	29	25	3
Other works	38	5	3	51	15	0
Bank protection	8	3	0	24	15	0
Embankment works	15	9	3	8	4	1
Miscellaneous works	18	5	1	22	3	1
Closure dyke	58	30	0	56	15	0

Table IV.2 and the two bar graphs (Figs. IV.1 and IV.2) show the distribution of costs for some dry basins and some wet ponds. Both the maximum and the minimum percentages have been given; the latter can be zero for some basins, indicating that neither equipment nor corresponding work existed in these cases; the most frequently occurring value has also been shown in the Table and labelled 'mode' (although it is not a mode in the statistical sense of the term).

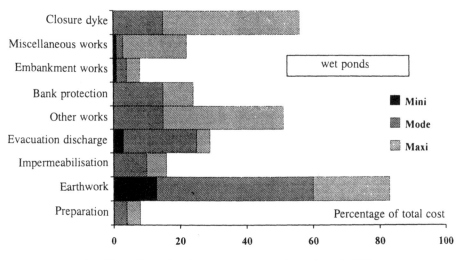

Fig. IV.1   Costs of various component works: wet ponds [76].

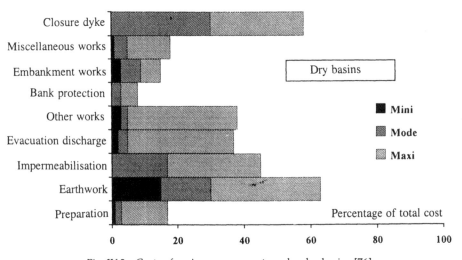

Fig. IV.2   Costs of various component works: dry basins [76].

The range of variation of percentages is obviously very large due to the small size of the sample—merely 21 basins constructed under very different conditions.

The maximal expenditure in almost all cases will fall under the heading 'earthwork'. It comprises the cost of bank fill and includes the cost of transport and laying of the material. The cost of removal of excess earth alone can double the cost of earthwork. That is why it is always preferable to use the on-site material in landscaping.

The cost of evacuation works is contributed more by the quantity of concrete used for civil works than by the price of mechanical equipment used for regulating the flow rate.

It may be noted that the costs of embankment works and protection of banks are relatively low—both for dry basins as well as wet ponds; the least expensive solution seems to be green areas.

The cost of impermeabilisation appears to be higher for dry basins than for wet ponds. This may be due to the particular sample analysed: of the seven dry basins studied, three involved construction works located close to major roads and consequently total impermeabilisation was necessary to protect the water table.

Lastly, it may be noted that the closure dyke, when provided, is the second largest heading under expenditure, for the simple reason that it involves very carefully executed earthwork, often requiring heavy machinery.

The cost of works related to (see Table IV.3):
— useful storage volume, effective surface area of the relevant catchment,
— surface area of the stormwater basin at the level N10 (10-year frequency level).

Costs were calculated in 1993 using the TP01 Index published every month in the 'Monitor of Building and Public Works' (France). It was found that as the volume, surface area of the basin, effective surface catchment area increase, the costs related to them decrease.

Calculations must include in the total capital investment the cost of collection conduits carrying water to the basin. Although one may be tempted to provide a single basin to handle flows from several collectors and to dimension this single basin, including the remediation works accordingly, the additional cost of the conduits required for this could well become higher than constructing several discrete basins (one for each conduit system).

Financial planners like to know the range of cost per m^3 of storage, as a function of the useful volume of the basin; it helps them determine the cost of retention for several possible flood situations.

As a matter of fact, the cost of a sewerage system depends much more on the volume of the retention basin (which determines the cost of earthwork and impermeabilisation ultimately) than on the complexity of regulation works. Nevertheless, it should be remembered that there is only a limited choice with respect to the outflow rate which essentially depends on the capacity of the outlet.

Urban planners are also interested in knowing the change in cost of basin per unit drained hectare as a function of the total effective drained area (this implies the effective catchment area). This helps in ascertaining the approximate cost of handling stormwater should a residential or commercial complex be built in this locale or on this surface subsequently.

Table IV.3 Per unit volume storage costs for the 7 dry basins and 14 wet ponds studied [76]

S. no. of stormwater basin		Invest- ment	V (ha)	Runoff coeff.	Effective surface of catchment area	Dry season bep	N10 bep	Useful vol. m³	bep-N10 BV effect	Free- board	Cost/m³	Cost/ha of effective catchment area	Cost/ha bep N10
3	S	2 282	109	0.26	28	0	1.20	73 000	4.2	2.9	31	81	1 902
7	S	8 262	230	0.79	182	0	2.80	70 000	1.5		118	45	2 951
8	S	4 275	50	0.65	33	0	0.35	13 000	1.1	3.8	329	132	12 214
11	S	11 833	155	0.42	65	0	1.50	70 000	2.3	5.5	169	182	7 889
18	S	3 769	250	0.23	58	0	0.30	13 000	0.5	5.0	290	66	12 563
19	S	5 347	300	0.25	75	0	0.50	22 000	0.7	< 8	243	71	10 694
20	S	1 902	10	0.25	3	0	0.16	3 000	6.4		634	761	11 888
1	E	5 476	115	0.30	35	1.30	1.50	70 000	4.3	1.1	78	159	3 651
2	E	15 255	61	0.70	43	2.00	2.00	70 000	4.7	0.5	218	357	7 628
4	E	2 399	37	0.67	25	1.95	2.30	25 350	9.3	0.5	95	97	1 043
5	E	3 226	50	0.43	22	1.00	1.50	22 700	7.0	0.7	142	150	2 151
6	E	9 742	133	0.37	49	2.60	2.60	67 000	5.3	0.7	145	198	3 747
9	E	11 672	920	0.29	267	5.50	5.50	80 000	2.1	0.9	146	44	2 122
10	E	18 121	1 300	0.33	429	3.63	4.00	60 000	0.9	1.5	302	42	4 530
12	E	3 934	11	0.65	7	0.50	0.60	6 500	8.8	0.8	605	576	6 557
13	E	32 418	467	0.49	229	2.58	4.00	380 000	1.7	3.0	85	142	8 105
14	E	5 300	10	0.60	6	0.50	0.60	5 500	10.0	0.7	964	883	8 883
15	E	9 995	587	0.49	288	7.00	7.00	210 000	2.4	2.0	48	35	1 428
16	E	13 908	424	0.56	237	2.50	4.00	240 000	1.7		58	59	3 477
17	E	7 292	177	0.32	57	0.25	0.50	30 000	0.9	5.0	243	129	14 584
21	E	5 962	135	0.65	88	3.38	3.60	44 800	4.1	0.7	133	68	1 656
1	2	3	4	5	6	7	8	9	10	11	12	13	14

(S dry basin; E: Wet pond; bep: surface area of stormwater basin; BV: total catchment area)
Effective surface of catchment area = Total catchment area × runoff coeff.

Investors are likewise interested in knowing the change in cost per hectare of basin as a function of the basin surface area at the once-in-ten-years flood level. This helps them estimate the additional costs arising from loss of land used for the basin which they then add to the cost of construction of lodgings, factories or zones of other activity. It was (and is) generally thought that stormwater basins occupy space which could more profitably be used for 'construction'. This is not always true since most basins are constructed in low-lying areas, even floodable land, or in zones specially reserved for them in urban master plans. Furthermore, the positive benefits accruing from protection of the environment and improvements in the quality of life should be taken into consideration when determining development costs. Building stormwater basins may appear at first glance to be underutilisation of the land but is in fact a revaluation or value added when the overalll interests of the community are taken into account.

IV-3.2.2   Underground tanks

The two headings that account for the maximum share of the total cost of an underground tank (costs of space acquisition not included) are the civil works (earthwork, concrete work) and equipment (pumps etc.).

The cost of civil works depend on:
— size of basin proposed (surface area, volume),
— depth,
— form of storage (shape),
— presence of groundwater aquifer(s) or other geotechnical constraints,
— number and nature of technical premises.

A study of twenty underground tanks, ranging in capacity from 1500 m³ to 65,000 m³ , revealed that the cost per cubic metre of storage varied from 500 to 6000 F (without taxes, 1993). Here, too, economy of scale was observed. The results are presented in Table IV.4.

As an example, Table IV.5 lists the cost of the Sevran basin in Seine-Saint-Denis (France) under various headings as a percentage of total cost.

IV-4   OPERATIONAL COSTS

The data available regarding actual cost of operation are very scant; thus only a few examples can be presented. This cost is generally lumped with other operational expenditures and often distributed among different units of management. Furthermore, remunerations to personnel are seldom revealed.

The cost of basin operation depends on the frequency of filling and the equipment used (pumps and machinery, automation, control system). For example, the annual expenditure related to the upkeep and maintenance of

Table IV.4   Cost of construction of some underground tanks

Town	Basin name	Volume (m³)	Cost MF (without taxes, 1993)	Cost/m³ of storage
Rennes	Cleunay	12,000	16,182	1,349
Le Havre		50,000		< 1,500
Nancy	Ducs-de-Bar	30,000	35,183	1,173
	Foch-Hardeval	18,000	29,023	1,612
	Gentilly	12,000	6,953	579
	Remicourt	11,000	7,830	712
	Vologne	1,500	5,742	3,828
	Marleville	13,000	10,962	843
	Haut-du-Lièvre	2,500	3,236	1,294
	Boudonville	3,400	3,863	1,136
Bordeaux	Perinot	39,000	35,490	910
	Maginot	25,000	23,660	946
	Migron-Bardenne	8,500	4,732	557
	RN 10	65,000	133,560	2,055
	Alhambra	12,000	19,627	1,636
	Bergonié	16,000	21,402	1,338
Seine-	Bangnolet	3,000	12,528	4,176
St. -Denis	Sevran	9,000	18,792	2,088
	Bondy	10,000	26,100	2,610
Vitry/Seine		60,000	170,172	2,836

Table IV.5   Distribution of total cost of Sevran basin, France (1993) among various headings of expenditure

Moulded walls	36%
Earthwork	12%
Slabs and columns	12%
Workshops/machine sheds	20%
Mechanical equipment	10%
Pumps and machinery	10%

equipment for control (excluding remunerations to personnel) lies between 0.50 and 4.00 F (without taxes, 1993) per cubic metre of storage.

The costs of removal of sludge and residues collected in the pretreatment works are not commonly available. Variation of these costs depends very much on the nature of the sludge and its eventual destination.

Some operational costs are common to both underground tanks and open ponds, for example, cost of sludge removal, maintenance of equipment for control and pretreatment. Others are more specific: cost of ecological maintenance of water bodies, maintenance of green areas, interventions in a confined environment. These differences are attributable to the significantly different remunerations to personnel of widely differing qualifications.

IV-4.1   Open Basins

To calculate the operational cost of hydraulic protection per se is difficult because costs related to other basin usages are often lumped together in the computation. Furthermore, an annual maintenance contract is often drawn up by the owner with a maintenance firm.

As for hydraulic usage, the calculation of cost should be limited to:

— maintenance of the storage capacity of basins (broadly speaking, removal of silt and other collected material);
— maintenance, in good running condition, of all instruments and equipment for control of flow rates (maintenance, cleaning);
— maintenance of equipment for treatment of effluents, both at the inlets and outlets of basins; the cost of routine analysis of water quality is subscribed under this heading;
— maintenance of the capacity of self-purification of the basin and, generally speaking, conservation of good 'ecological' conditions.

Expenses under the first three headings can be computed exactly from the account books by including the actual costs of items for programmed repairs and replacements; the only condition is that the cost of basin operation be not included in calculation pertaining to maintenance of infrastructure or roads etc.

Estimation of expenditure under the fourth heading is difficult. In principle, it includes the cost of usual interventions during the year (cleaning, plantations, repair of damaged banks) and the shared cost of periodic operations such as emptying, remediation, weeding, stocking fish for maintenance of ecological equilibrium etc. The cost of ecological upkeep and related biological analysis should also be included.

Rosenbaum [68] has cited some costs for the usual operations:

— Routine maintenance of a wet pond, including weekly collection of flotsam, plant control (annual weeding, removal and transport of cuttings to the dump), maintenance of screens and operation of various valves may cost 30 to 40 F (without taxes, 1993) per running metre of bank. This cost can significantly increase with distance to the dump. Agricultural recycling by composting the products of weeding can reduce this cost.
— Monitoring the water quality by physicochemical analysis at the rate of two tests per annum and one sample from every 4 hectares of water body may cost 5000 to 15,000 F (without taxes, 1993) for each sampling site.

In constructed stormwater basins maintained as natural ponds, the cost of annual maintenance varies between 20,000 and 50,000 F per hectare of water body, and the cost of ecological upkeep between 1500 and 4000 F per hectare (cost without taxes, 1993 projects)

The costs of maintenance of dry basins correspond to those of maintenance of green areas, collection of damaging objects and detritus brought in by rain, cleaning the central trench and maintenance of control works and equipment.

IV-4.2 Underground Tanks

In contrast to open basins, the costs of operation and maintenance of underground tanks depends on the degree of automation of devices for cleaning the tank. Here, evacuation in most cases has to be done by pumping. Thus the cost of pump wear and power consumed has also to be taken into account.

For example, the annual cost of manual cleaning of the underground tank of Montigny (Sivom de Metz) came to 25,000 F (without taxes, 1993) inclusive only of personnel remuneration. Maintenance was carried out every two months to remove deposited silt.

In several underground tanks in the Aquitaine region (France), the cost of cleaning is known to be 5 to 7 F (without taxes)/m^2 (base 1992, 4 to 8 operations per year).

5

LEGAL REGULATIONS (IN FRANCE)

V-1 WATER AND ENVIRONMENT

Regulations in France regarding the protection of natural resources and, in a broader sense the environment, have considerably evolved between 1964, the year in which the first law on water was promulgated, and the publication of this guide.

V-1.1 Stormwater Control

Law 92-3 of 3 January 1992, article 35.III requires urban communes to delimit, after a public investigation:
— zones in which measures should be taken to limit the impermeabilisation of land and to ensure stormwater control,
— zones in which it is necessary to provide devices for collection, storage and when needed, treatment of stormwater if the pollutant load discharged by it to the receiver waters is likely to seriously damage the efficiency of the sewerage system.
The above two paras of article 35.III invoke directly or indirectly the need for retention basins which may form part of the array of devices recommended for good management of the water cycle.
The above provisions relating to zoning were immediately applicable.

V-1.2 Discharge Permit

The bylaw of 20 November 1979, framed for application of decree no. 73-218, lays down provisos applicable to the stormwater discharged from channels, pipes (decree no. 73-218 was later abrogated).
Law 92-3 of 3 January 1992 prescribes in article 10 the procedures for declaration or permit for works involving pairing out, flow, discharge or storage which may be direct or indirect, chronic or occasional, polluting or non-polluting.

Decree no. 93-743 of 29 March 1993 gives the list of these works in its annexure, notably:

Clause 2.7.0

Creation of ponds or water bodies with surface area:

1) larger than 3 ha would need Permit;
2) larger than 2000 m^2 but smaller than 3 ha would need Declaration.

Clause 5.3.0

Regarding discharge of stormwater into surface waters or in an infiltration pond if the total water catchment area is

1) larger than or equal to 20 ha, a Permit is required;
2) larger than 1 ha but smaller than 20 ha, a Declaration is required.

V-1.3 Impact Study (Environmental Impact Study or Statement)

Article 2 of law 76-629 of 10 July 1976 relating to protection of nature states:

The preliminary study before planning or installation of works which, by virtue of their size or their influence on the natural environment, could undermine the latter, should include an impact study which would help in assessing the consequences.

Decree no. 77-1141 of 12 October 1977, framed for implementation of article 2 of the above law, specifies the works exempted from the impact study. In particular, it exempts 'the underground or semi-underground reservoirs for water storage'.

Thus only open basins, irrespective of their size, are subject to an impact study under current French law.

V-1.4 Policing of Fishing

Article L.231-3 of the French Rural Code, whose modalities of application are specified in the circular on environment of 16 September 1987, stipulates that the works creating water bodies, connected even in a discontinuous manner with a watercourse, are subject to policing of fishing.

V-2 DAMS

The decree of 13 June 1966 constituted a Permanent Technical Committee for Dams; article 2 of this decree reads:

...Irrespective of who the owner is, the Permanent Technical Committee for dams must be obligatorily consulted by the ministry concerned with respect to the preliminary as well as the final design of dams more than 20 metres higher than the lowest point of the undisturbed ground.

It may be recalled that stormwater retention basins do not generally require construction of dykes greater than 20 metres in height.

The circular of the French Ministry of Agriculture dated 22 October 1974 relates to dams of small or moderate height constructed on watercourses not belonging to state and those constructed away from watercourses and distinguishes between the two. The circular further specifies that the dams constructed away from watercourses are not subject to the control of the 'Water Police'. They are, however, governed by the interdepartmental* circular of 14 August 1970 concerning the inspection and supervision of dams from a public safety point of view. **Under the provisos of this circular, the DDAF or DDE** can include a retention basin in the list of dams/ reservoirs involving public safety.**

V-3   PUBLIC SAFETY AND HEALTH

As they are reservoirs of water not meant for drinking, the retention basins are governed by the municipal police as specified in the Code of Communes and Departmental Sanitary Regulations.

V-3.1   Powers and Responsibilities of the Mayor

Article L.131-2 stipulates:
> **The mayor is responsible...for taking suitable precautions to put a stop to pollution of all types, such as that caused by flooding or breaking of dykes, by dispatching required help.**

Article L. 131-11 stipulates:
> The mayor can require the owners, beneficiaries and users to enclose by proper fencing, all holes, wells, shafts and excavations likely to endanger public safety.

V-3.2   Departmental Sanitary Regulations

Article 36 of the Standard Departmental Sanitary Regulations concerning storage of non-potable water stipulates:
> Reservoirs of non-potable water, beautification ponds, those for irriga- tion as well as all other water storages have to be emptied as often as necessary, in particular to prevent proliferation of insects. They have to be cleaned and disinfected as often as necessary, but at least once a year.

---

*France is presently divided into 95 departments (plus 4 overseas) equivalent to countries in U.K.

**Public department in charge of control of watercourses

These provisos may not be strictly applicable to stormwater retention basins.

V-4 URBAN PLANNING

Retention basins have not been explicitly mentioned in the Code for Town Planning. They may, however, be considered a part of lowering or raising the level of land operations, which are frequently mentioned in the Code.

V-4.1 General Principles of Town and Country Planning

Article L.121-10 states:
"The town planning documents specify conditions that assist in prevention of natural as well as technological risks. The provisos of this article have to be considered as law for Town and Country Planning in the sense of article L.111-11 of the French Code of Urbanisation."

This article confers a special importance on the articulation between elaboration of town planning documents and studies of stormwater networks and retention basins. It refers to the obligations of the communities regarding division of their territory into zones (cf. V-I.1) and the measures limiting the impermeabilisation of land conforming to law 92-3 of 3 January 1992.

Article L.123-1 reads:
'The POS* should fix the sites reserved for public roads and works....

Obviously, retention basins should figure among the sites reserved by POS.

In this context, attention is drawn to the consequences of article L.123-9 which specifies the rights of the owners and the obligations of the community with respect to reserved sites.

V-4.2 Elaboration of POS

Retention basins are governed by the following articles:
R.123-18: Graphic documents
"The graphic documents show, where relevant, the entire zone or a part thereof in which the requirements of the operation of public services for health, protection against nuisances...justify the prohibition or subjection to special conditions of all types of buildings and facilities, permanent or otherwise; plantations, depots, lowering, drilling and raising of land...; sites reserved for public roads and works."

---

*Plan d'Occupation des Sols: Land Use Plan. This refers to documents which define the way each part of the community territory has to be used.

R.123–21: Regulation
This regulation specifies the rules applicable to the terrains included in various zones of the territory covered by the plan.
(1) For this purpose:
a) the major allotments of sites should be specified by zones in accordance with the categories provided in R.123–18 by prescribing the principal usage to which they can be put and, if necessary...the operations of lowering and raising the land.

V-4.3 Building Permit

According to the Code of Urbanisation, a request for building permit must be made in most cases of retention basins.
Article L.421–1 reads:
Anyone desiring to undertake construction for residence or otherwise, even without foundations, must first obtain a building permit, according to the provisos of articles L.422–1 to 422–5.' This requirement is also binding on the public services and the concessionaires of public services of the state, region, department or commune as well as on private bodies (French law 86–13 of 6 January 1986, article 2–II).
This general scheme of building permits is the subject of derogation whose conditions are specified in articles R.421–1 and R.422–1:
Article R.422–1 reads:
...the following works or installations are not included in the scope of application of the provisos of the building permit:
(1) Underground works or installations of gas or fluids storage....
The underground retention tanks are thus exempted.
Article R.422–2 reads:
"The following are exempted from the requirement of building permit:
(h) Among the technical installations required for the functioning of public services in charge of drinking water supply and sewerage disposal, the technical structures whose floor area is smaller than 20 square metres and height less than 3 metres....
(k) Uncovered swimming pools...."
Note: French Circular DAU* of 25 July 1986 contains information that does not coincide with the contents of the above articles of the Code of Urbanisation.
The revised article R.421–1 of the Code of Urbanisation gives, under the heading 'especially' a non-exclusive list of works or structures exempted from the application of building permit. Before focussing attention on the *procedure of authorisation or declaration*, it would be a good idea to examine the complementatry provisions contained in the

---

*Direction de l'Architecture et de l'Urbanisme of Ministry of Public Works and Transport

doctrine or the *jurisprudence*. This would lead to the conclusion that the following should continue to be considered excluded from the scope of application of the building permit....:

(3) Basins, irrespective of their usage: pleasure, agricultural, pisciculture, aquaculture etc. (and) uncovered swimming pools already covered by the exemption list.

V-4.4 Preliminary Declaration

Article L.422–2 stipulates that the works exempted from the building permit should nevertheless be subject to a preliminary declaration.
Article L.422–2 reads:

The constructions or works exempted from the building permit, except for those covered by the secrecy of national defence, are subject to making a declaration before the mayor before starting the work.

Article R.422–3 reads:

...The declaration includes the identity of the declarant, the location and area of land, owner's identity (if he is not the declarant), the nature and purpose of the works and if relevant, the density of the constructions, both existing and those to be created.

The file attached to the declaration shall include a map showing the location of the land and a graphical document in which buildings and their surroundings are schematically drawn. This document shall show external outlines, traffic ways, tree plantation etc. and shall include an overall plan, a block plan, a layout plan and a three-dimensional development plan.

V-4.5 Authorisation of Various Installations and Works

According to the French Code of Urbanisation, retention basins belong to the group of various installations and works for which a request for authorisation must be made.
Article R.442–2 reads:

In the communes or parts of communes mentioned in article 442–1...before commencing construction and/or installation of works covered by the categories enumerated below, a preliminary authorisation for the same has to be obtained.... (c) Works involving lowering or raising of land when the surface area is more than 100 m^2 and the height in the case of raising and the depth in the case of lowering exceeds two metres.

V-4.6 Financing of Works

The Code of Urbanisation provides that private operators may be required to meet all or at least part of the cost of public facilities needed for the project; retention basins may be covered by this proviso.

L.332–8 reads:

> The beneficiaries of authorisation may be required to share the cost of all industrial, agricultural, commercial or artesanal installations which by virtue of their nature, location or importance necessitate unusual acquisition of public facilities.

L.332–9 (PAE)[†] reads:

> In the areas from the 'commune' territory where town and country planning was approved by the muncipal council, the council can decide that the building permit beneficiary is responsible for creating all or part of public facilities needed for present and future inhabitants of the area in accordance with the town and country plans.

V-4.7　Construction of Retention Basins in the NC* and ND** Zones

The NC and ND zones are defined in sections (c) and (d) of Article R.123–18 of the Code of Urbanisation. These definitions may lead to questions regarding the construction of retention basins in these zones. It would be a good idea to include the construction of basins in the definition of NC and ND zones.

V-5　ACCIDENTAL POLLUTION

The law titled 'fishing', no. 84–512 of 29 June 1984, and the law on water, no. 92–3 of 3 January 1992 include essential legal means for initiating judicial action in the case of accidental pollution. In the case of a 'classified' installation, law 76–663 of 19 July 1976 titled 'classified installations' describes the additional penal and administrative sanctions applicable, the details of which are not given here.

Article L.232–2 of the Rural Code establishes the offence of pollution:

> Anyone who pours or lets flow into water as mentioned in article L.231–3 of the Rural Code, directly or indirectly, substances whose action or reaction is destructive for fish or harmful to their nutrition, reproduction or alimentary function, shall be punished....

Article 22 of the law on water dated 3 January 1992 extends the concept of offence of pollution by taking into account the ensemble of water

---

†PAE: Plan d'Aménagement dénsemble. A contract between the community and builder about integration of a newly constructed area in the city.

*NC zones are protected areas with respect to their agricultural value, soil fertility, underground resources etc.

**ND zones are protected areas with respect to their high aesthetical, historical and ecological interests. They can also be protected because of risks of nuisances.

resources and all 'harmful effects on health or damage to the flora and fauna, excluding the damages covered in article L.232–2 of the Rural Code...or significant modification of the normal regime of the food chain in the water body or limitation of usage of the bathing zones'.

Article 18 of the law of 3 January 1992 stipulates that if a person has knowledge of any incident posing a threat to civil safety, the quality, supply or conservation of water, he is obligated to communicate this information to the concerned prefect or mayor. It also obliges the person responsible for generating the pollution to take necessary measures to eliminate its cause and spread to the aquatic medium. It further empowers the concerned authority to take remedial measures at the expense of the person responsible for the pollution. Furthermore, the prefect and the mayor are duty-bound to keep the public well informed about any such incident.

When an accidental pollution is reported, a statement has to be recorded which implies that a violation has been committed. In the case of renewed pollution or one that persists for several days, successive statements must be filed. The competent authority for recording such statements is defined in article 2 of the law of 3 January 1992 and in article L.237–1 of the Rural Code under the heading 'law of fishing'.

The civil legal process is independent of any penal action taken by the water police and may be pursued even in the absence of police action; its objective is to obtain compensation for the victims and is based on articles 1382, 1383, 1384 al. 1 of the Civil Code.

Case Studies

Representative cases of two types are included here: ensembles of retention basins and isolated basins. The first type presents the arrangement of basins to demonstrate the general operation of the system. The associated catchment area and the regulated flow rate are given for each stormwater basin. The description has purposely been kept schematic. The system of detention ponds situated along Maubuée stream (Ville Nouvelle de Marne-la-Vallée) and that of dry basins of Vitrolles (La Cadière stream) are described as representatives of the first type. For the second type, all basic information available on basins constructed in France is presented including that on the more common ones.

Financial information relates either to the observed costs updated as per the Public Works Index or estimates arrived at during updates of the APD.* Institutional information may be incomplete or altogether missing, especially for older basins.

---

*Avant Projet Détaillé: Detailed Project Study

SYSTEM OF STORMWATER BASINS AT MAUBUEE STREAM
(MARNE-LA-VALLEE)

Ville Nouvelle de Marne-la-Vallée (Seine and Marne Rivers)

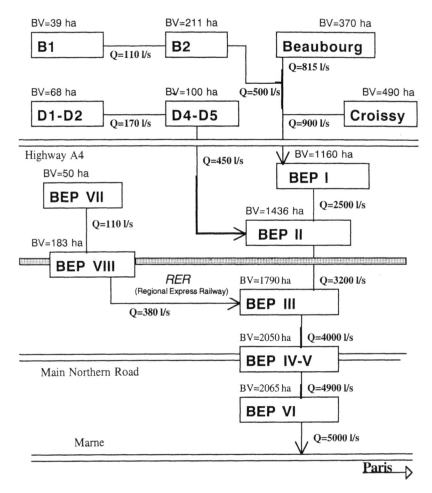

The system of stormwater management of Sector II of Ville Nouvelle de Marne-la-Vallée consists of an ensemble of 13 detention ponds.

The limiting factor in this case was the rate of outflow before reaching the Marne River: 5 m³/s. If no protection had been provided during the urbanisation of this sector, the flow rate of a ten-year flood situation in Maubuée as delivered at the Marne would have exceeded 50 m³/s (the average low flow rate is almost 25 m³/s).

Most of the basins have added value by providing landscapes of high quality and their integration with the urban environment is excellent. Water flowing into the ensemble is of a lovely sea-green colour; it debuts from the old Croissy pond (which now receives the stormwater of the urbanised area) and terminates in the new VI basin.

Basin VII (photo below) well demonstrates successful urban integration. The flow control equipment is hidden by a stairway whose steps act as weirs.

DETENTION POND OF MANDINET (basin VIII)

Ville Nouvelle de Marne-la-Vallée (Seine and Marne Rivers)

The Southern and Northern subbasins of basin VIII belong to the large ensemble of basins assuring regulation of the stormwater reaching Maubuée stream.

The Southern subbasin of basin VIII presents a particularly unique type of urban integration. The railway line cuts across the basin and the escalators of Lognes-Mandinet station lead to banks formed as quays. A weir shaped like an inclined stone wall connects the Southern and Northern subbasins of basin VIII.

The two subbasins form an attractive water body, encircled by a promenade, and foster a duck population comprising several species. Anglers have also found a haven here.

Stormwater basin   **ZAC DU MANDINET (basin VIII)**
Institutional data

Owner:	SAN Val Maubuée, EPA Marne
Project supervisor	SAUVETERRE and DDE 77
Operator:	SFDE

Financial data

Year of construction: 1979	Duration of construction work:	1 year
Total cost (Francs, 1993, without taxes):	9.7 MF	
Part contributing to flood protection:		100%
Part specially provided for ancillary uses:–		

Technical data

Relevant water catchment area: 180 ha   Surface area of basin: 3 ha
Runoff coefficient: 0.37   Storage volume: 67,000 m^3
Drawdown range 10 years: 68 cm   Drawdown range 100 years: 170 cm
Flow-control device: calibrated orifice
Regulated flow rate: 380 l/s
Phreatic (free) groundwater: yes   Watertightness: yes
Receiving waters: Maubuée stream
Pretreatment: Screen and grit chamber
Dyke: yes
Special features: exemplary consultation among city planning and sanitary engineers, resulting in landscaping of exceptional quality
Major problems encountered: some pollution by inflow of city waste waters

DETENTION POND OF REGIONAL URBAN CENTRE

Commune of Noisy-le-Grand (Seine-Saint-Denis)

This basin of 2 hectares is situated in the commercial and residential zone of Mont d'Est. It is fed by stormwater from Pavé street as well as drainage of the 4-km long open trench for the Express Regional Railway.

This lined basin, well integrated from the architectural point of view, incorporates waterlily flotillas and gushing fountains and gives the feeling of a miniature natural lake. But the occasional inflow of contaminated water and a large population of ducks have posed problems. The water jets and fountains help in oxygenation of water. Regular ecological upkeep of the basin since its construction has proved highly beneficial.

Stormwater basin **REGIONAL URBAN CENTRE**
Institutional data

Owner: EPA Marne
Project supervisor: BETURE-SETAME
Operator: Technical Service, Noisy-le-Grand

Financial data

Year of construction: 1977   Duration of construction work: 1 year
Total cost (Francs, 1993, without taxes): 13.8 MF
Part contributing to flood protection: 100%
Part specially provided for ancillary uses: –

Technical data

Relevant catchment area: 61 ha   Surface area of basin: 2 ha
Runoff coefficient: 0.7   Storage volume: 70,000 m^3
Drawdown range 10 years: 54 cm   Mean depth: 100 cm
Flow-control device: calibrated orifice   Regulated flow rate: 800 l/s
Phreatic (free) groundwater: yes   Watertightness: yes
Receiving waters: storm sewer
Pretreatment: none; only aeration by cross-current jets
Dyke: no
Special features: sophisticated landscaping; this basin has attracted both residential and commercial interests
Major problems encountered: some parasitic pollution by waste waters. Self-purification capacity low because vegetation sparse and non-diversified

DETENTION POND Marne-la-Vallée no. 18, Sector IV

Commune of Bailly-Romainvilliers (Seine and Marne Rivers)

This basin regulates stormwater from a camping site and proposed future urbanisations. It is built on the Folie stream, which transits a grit chamber before entering the basin. A flow-breaker is hidden under a wooden footbridge.

An island stretches along the linear length of the bank to facilitate better plant growth and provide shade for ducks and other waterfowl that have colonised in spite of its proximity to a very busy road.

Stormwater basin   **MARNE-LA-VALLEE no. 18, Sector IV**
Institutional data

Owner:	SAN des Portes de la Brie-EPA France
Project supervisor:	Sauveterre, DDE 77, TUGEC
Operator:	SAN des Portes de la Brie

Financial data

Year of construction: 1990   Duration of construction work: 1 year
Total cost (Francs, 1993, without taxes): 6.3 MF
Part contributing to flood protection: 100%
Part specially provided for ancillary uses:–

Technical data

Relevant catchment area: 150 ha   Surface area of basin: 3.7 ha
Runoff coefficient: 0.39   Storage volume: 56,000 m^3
Drawdown range 10 years: 50 cm   Drawdown range 100 years: 150 cm
Flow-control device: shutter with adjustable sectional area (ITERA make)
Regulated flow rate: 980 l/s
Phreatic (free) groundwater: yes   Watertightness: no
Receiving waters: Folie stream
Pretreatment: Grit chamber at inlet to basin
Dyke: flow-breaker after the grit chamber provided by road embankment
Special features: real-estate potential for future urbanisation of the sector made more attractive and integrated for increasing surface area of basin
Major problems encountered: none to date

DETENTION POND OF COURANCE

Commune of Maurepas (Yvelines)

The retention basin of Maurepas-Courance is one part of the plan of development of the hydrographic network of the plateau of Trappes-Saint-Quentin-en-Yvelines following intense urbanisation of the region. There are some fifteen retention basins in this region and the height of their dykes varies between 3 and 16 m. The basin of Maurepas is one of the largest (capacity-wise) with a high dyke. It has been developed for fishing and promenades in a site surrounded by a natural forest.

The highest dyke is 15.50 m and its length is 245 m. This dyke is equipped with a surveillance system and is regularly inspected.

Stormwater basin   **MAUREPAS-Courance**

Institutional data

Owner:	SAN de Saint-Quentin-en-Yvelines
Project supervisor:	DDAF
Operator:	Commune (Technical Department)

Financial data

Year of construction:	1974-77	Duration of construction work:	4 years
Total cost (Francs, 1993, without taxes):		48 MF	

Part contributing to flood protection:–
Part specially provided for ancillary uses:–

Technical data

Relevant catchment area:	900 ha	Surface area of basin:	8 ha
Drawdown range 10 years:	400 cm	Storage volume:	393,000 m^3
Flow-control device:	valve	Regulated flow rate:	2.3 m^3/s
Phreatic (free) groundwater:	yes	Watertightness:	no
Receiving waters:	storm drainage, Courance		

Pretreatment: screen, grit chamber and oil and grease removal unit

Dyke:	yes	Watertightness:	no

Special features: fed by underground aquifer and seepage drain of Courance

Major problems encountered: during construction lowering of water table and extraction of peat had to be undertaken

DETENTION POND OF ZAC DES LONGS CHAMPS

Commune of Cesson-Sévigné (Ille-et-Vilaine)

ZAC des Longs Champs was constructed in the proximity of the complex of Beaulieu University in eastern Rennes.

The rate of outflow allowable was only 300 l/s while urbanisation was likely to generate a flow rate of 10-year flooding of the order of 6000 l/s. One basin existed already but its capacity was inadequate. A second basin was constructed and is an example of integrated landscaping as much for the basin and its dyke as for the works installed in the faces of a retaining wall and approached by a monumental stairway leading to the basin.

Front view of the loophole-shaped regulation works and the spillway weir for safety against flooding for more than a hundred years.

Storm run-off basin   ZAC DES LONGS CHAMPS
Institutional data

Owner:	SEMAEB
Project supervisor:	Sauveterre, Ville de Rennes
Operator:	Ville de Rennes, Technical services

Financial data

Year of construction: 1979   Duration of construction work: 4 months
Total cost (Francs, 1993, without taxes): 6.8 MF
Part contributing to flood protection: 76%
Part specially provided for ancillary uses: 24%

Technical data

Relevant catchment area: 95 ha   Surface area of basin: 2.9 ha
Runoff coefficient: 0.48   Storage volume: 42,000 m^3
Drawdown range 10 years: 50 cm   Mean depth: m
Flow-control device: AVIO valve 71/40 + module with mask XXI
Regulated flow rate: 300 l/s, of which 225 l/s for floods
Phreatic (free) groundwater: yes   Watertightness: partial
Receiving waters: stream
Pretreatment: none
Dyke: yes: made of rubble extracted from the basin
Special features: landscaped basin and associated greenery has become a strong point of the residential area, animated by a rich fauna of birds, especially ducks
Major problems encountered: none to date

DETENTION POND OF FONTAUDIN

Commune of Pessac (Gironde)

Designed for regulating stormwater from three subbasins (surface area about 300 ha), this work is part of the programme of general development of the catchment area of Ars stream which covers 2300 ha spread over the communes of Pessac, Talence, Begles and Bordeaux.

It was designed with a view to preserving the environment as much with respect to the water quality as development of landscaping and preclusion of any type of nuisance; hence pretreatment of water, landscaping the dykes and the embankment with gentle slope and turfing, and construction of a building with suitable architecture for locating the pretreatment and regulation units.

Apart from its hydraulic role, the basin is used for fishing, promenade and picnics.

Stormwater basin   FONTAUDIN

Institutional data

Owner: Urban community of Bordeaux (CUB)
Project supervisor: DDE Gironde
Operator: Lyonnaise des Eaux-Dumez

Financial data

Year of construction: 1984   Duration of construction work: 1 year
Total cost (Francs, 1993, without taxes): 12.5 MF
Part contributing to flood protection: –
Part specially provided for ancillary uses: –

Technical data

Associated catchment area: 300 ha   Surface area of basin: 3.2 ha
                                      Storage volume: 48,000 m^3
Drawdown range 10 years: 150-200 cm
Flow-control device remote controlled valves   Regulated flow rate: 1.2 m^3/s
Phreatic (free) groundwater: yes   Watertightness: no
Receiving waters: storm drainage network, then Serpent River
Pretreatment: screen, oil and grease remover, grit chamber
Dyke: no
Special features: fed by water table
Major problems encountered: the association of residents was opposed to this project. However, they began to use this public space after it was integrated into a landscaped park

SYSTEM OF STORMWATER MANAGEMENT BY
BASINS OF VITROLLES
Commune de Vitrolles (Months of Rhône river)

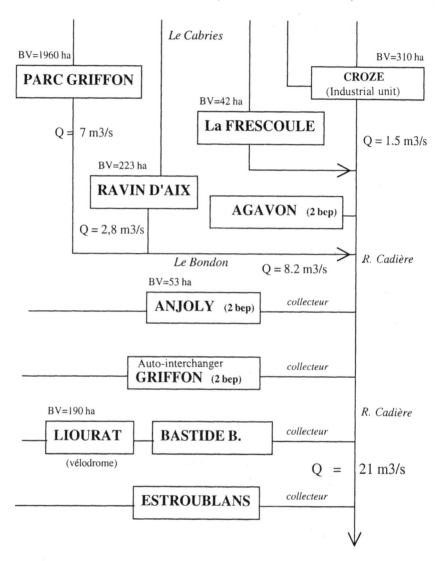

The drainage system of Ville Nouvelle de Vitrolles now comprises 13 dry basins. This arrangement enables control of the discharge of the Cadière river to 21 m³/s during flooding (roughly 7 l/s/ha), thereby preventing inundation in the region of the airport of Marseille-Provence and the towns of Marignane and Estoublet located downstream.

The rate of outflow is relatively higher due to the fact that in Mediterranean climate the amount of stormwater is larger. The basins empty very fast and the duration of non-availability for other usages is generally only a few hours after a storm episode.

The photo below shows Griffon Park. It is in fact a dry basin of surface area 4 ha and volume 56,000 m^3, which is used as an attractive park and playground outside the storm period.

DRY BASIN OF FRESCOULE

Commune of Vitrolles (Bouches du Rhône Departement)

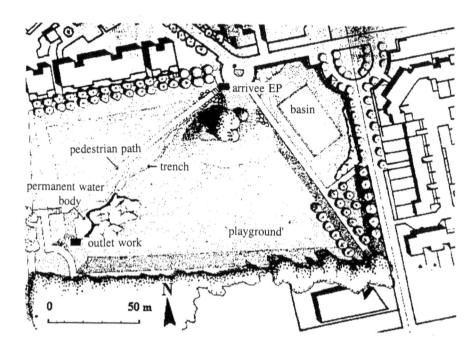

(Document EPAREB)

The basin of Frescoule (a neighbourhood with 900 housing units) belongs to the ensemble of retention basins located in the catchment area of Cadière river. It comprises, in fact, two basins: one rectangular and located on an upland to the north-east and used for making small-size model boats. A lower level area, consisting of a large space used as a 'playground' and a public park, holds a second, smaller permanent basin that is very aesthetic, filled with greenery and fed by a stream at the south-west corner. There is an adjoining paved pedestrian walkway. The ensemble, including a kiosk, forms a panoramic view and is an example of successful combination of architecture and landscaping (see colour photo 10). It may be noted that the inlet of rainwater is concealed by a rock-garden; the technical installations are also concealed.

Stormwater basin   **FRESCOULE**

Institutional data

Owner: Commune de Vitrolles, EPAREB
Project supervisor: Society du Canal de Provence
Operator: Commune de Vitrolles, Technical Services
Financial data
Year of construction: 1984   Duration of construction work: 6 months
Total cost (Francs, 1993, without taxes): 5.7 MF
Part contributing to flood protection: 50%
Part specially provided for ancillary uses: 50%

Technical data

Associated catchment area: 120 ha   Surface area of basin: 1.5 ha
                                       Storage volume: 25,000 m^3
Drawdown range 10 years: 180 cm
Flow-control device: gate valve
Equation governing evacuation: $Q = 0.385\ H$ (m^3/s) (H in metres)
Phreatic (free) groundwater: yes   Watertightness: no
Receiving waters: cross-stream
Pretreatment: none; though there is some daily pollution in the cross-stream by an agro-food products industry (washing of potatoes)
Dyke: no
Special features: very carefully executed landscaping. A didactic notice displayed at the entry reminds the public that the work plays primarily a hydraulic role of toning down the effect of floods.
Major problems encountered: pollution in the cross-stream which flows into the wet pond

DRY BASIN OF LIOURAT

Commune of Vitrolles (mouths of Rhône River)

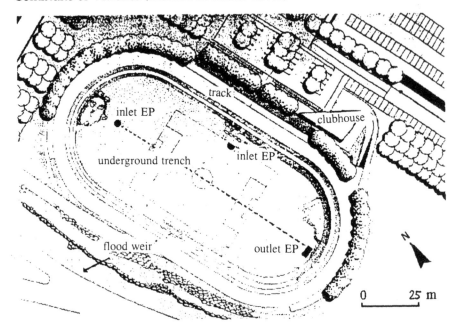

inlet EP
track
clubhouse
underground trench
inlet EP
flood weir
outlet EP
N
0   25 m

(Document EPAREB)

The basin of Liourat also belongs to the ensemble of retention basins located on the catchment of Cadière River.

The economic exploitation of the rubble quarry for building the embankment of the French Railway is worth noting. The basin is multifunctional: hydraulic role and playgrounds: velodrome and stadium (see colour photo 5).

Water from light rain flows through the bypass situated under the velodrome so that the playgrounds are not unnecessarily inundated. The velodrome is useable throughout the year except for a few hours during heavy storms.

Stormwater basin **LIOURAT**

Institutional data

Owner: Commune of Vitrolles, MOD: EPAREB
Project supervisor: SCP, SCOP AB2i
Operator: Sports Service of Vitrolles city

Financial data

Year of construction: 1985   Duration of construction work: 9 months
Total cost (Francs, 1993, without taxes): 9.4 MF
Part contributing to flood protection: 69%
Part specially provided for ancillary uses: 31%

Technical data

Associated catchment area: –   Surface area of basin: 1.4 ha
     Storage volume: 40,000 m^3
Mean depth near approach to weir: 2.8 m
Flow-control device: circular gate valve
Equation governing evacuation: $Q = 0.46\ H\ (m^3/s)$ (H in metres)
Phreatic (free) groundwater: no   Watertightness: no
Receiving waters: cemented drain followed by stream
Pretreatment: none
Dyke: no, but approach of weir protected by rocks
Special features: abandoned quarry converted into a stormwater basin with velodrome and stadium
Major problems encountered: maintenance of approach, lawns and plants

DRY BASIN OF CHEMIN DE CLERES

Commune of Bois-Guillaume (Seine Maritime)

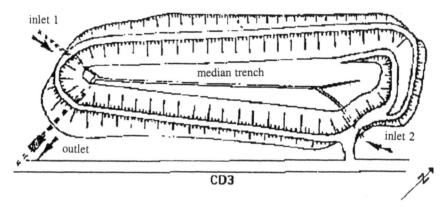

The dry basin of Chemin de Clères, close to Rouen, has an exceptional functional characteristic. It has no other usage except hydraulic protection. But great care has been taken in regard to the appearance of the lawn. The basin is prismatic in shape and a median trench at the bottom evacuates small inflows from the inlets (see colour photo 7).

Stormwater basin   **CHEMIN DE CLERES**

Institutional data

Owner:	S.I.A.R.R.
Project Supervisor:	DDE-76
Operator:	S.I.A.R.R.

Financial data

Year of construction: 1977   Duration of construction work: 1 year
Total cost (Francs, 1993, without taxes): 4.5 MF
Part contributing to flood protection: 100%
Part specially provided for ancillary uses: –

Technical data

Relevant water catchment areas: 250 ha   Surface area of basin: 3000 m^2
Runoff coefficient: 0.23   Storage volume: 13,000 m^3
Drawdown range 10 years: –   Mean depth: 5 m
Flow-control device: calibrated orifice   Regulated flow rate: 300 l/s
Phreatic (free) groundwater: yes   Watertightness: yes
Receiving waters: stream
Pretreatment: none; flow-breaker and trench for small flows
Dyke: yes; made partially of rubble dug from the basin
Special features:–
Major problems encountered:–

DRY BASIN OF COUBRON COMPARTMENT

Commune of Coubron (Seine-Saint-Denis)

The purpose of Coubron basin is to combat flooding and pollution of Chantereine stream before it empties into the Marne river. The basin has been integrated in a pavilion-type urban fabric; its landscaping was executed with great care.

The basin comprises the following two zones whose levels differ by one metre:

— a covered sedimentation basin , volume 8500 m^3,
— a paved zone prone to inundation, volume 10,000 m^3.

The stormwatern flows into the basin by gravity but needs to be evacuated by pumps.

Stormwater basin   **COUBRON**
Institutional data

Owner: Council General of Seine-Saint-Denis
Project supervisor: Directorate of Water and Sanitation
Operator: Council General

Financial data

Year of construction: 1989-90   Duration of construction work: 2 years
Total cost (Francs, 1993, without taxes): 19.5 MF
Part contributing to flood protection: 50%
Part specially provided for ancillary uses: 50%

Technical data

Associated catchment area: 170 ha	Surface area of basin: 1.5 ha
Effective volume 10 years: 20,000 m^3	Storage volume: > 30,000 m^3
Drawdown range 10 years:–	Mean depth:–
Flow-control device: pumps	Regulated flow rate: 1 m^3/s
Phreatic (free) groundwater: yes	Watertightness: yes

Receiving waters: Chantereine channel
Pretreatment: sedimentation in the first chamber
Dyke: no
Special features: fed by two sewer drains
Major problems encountered: obligation of drainage being done only after a detailed study to ensure protection of downstream ponds

DRY BASIN OF LEYSOTTE

Commune of Villeneuve d'Ornon (Gironde)

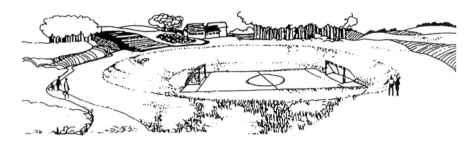

The objective of this public space prone to inundation is to restrict the large amount of stormwater generated during floods that reaches the inadequate drainage collectors downstream. To simultaneously upgrade the quality of development of the dense urban area (consisting mostly of residential plots), complementary usage through sports activity was also planned.

This basin, constructed on a natural clay terrain, is located on a diversion of a stormwater sewer of 1000 mm diameter, which feeds it by overflow.

Stormwater basin   **LEYSOTTE**

Institutional data

Owner: Urban Community of Bordeaux (CUB)
Project supervisor: Lyonnaise des Eaux-Dumez
Operator: Lyonnaise des Eaux-Dumez

Financial data

Year of construction: 1986   Duration of construction work: 1 year
Total cost (Francs, 1993, without taxes): 2.7 MF
Part contributing to hydraulic protection: –
Part specially provided for ancillary uses: –

Technical data

Associated catchment area:–   Surface area of basin: 0.7 ha
        Storage volume: 7200 m^3
Flow-control device: valves and gates   Regulated flow rate:–
Phreatic (free) groundwater: no   Watertightness: no
Receiving waters: Stormwater network
Pretreatment: screens
Dyke: no
Special features: dual-purpose: hydraulic protection and football field
Major problems encountered: cleaning needed after every filling; regular mowing required

---

DRY BASIN OF BEQUIGNEAUX

Commune of Bruges (Gironde)

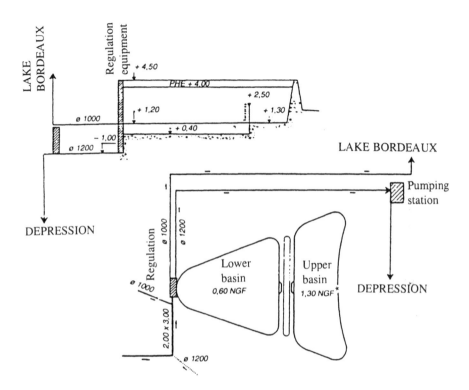

This basin was constructed to protect the downstream drainage network from overflow caused by heavy rains of short duration and to divert the most polluted stormwater towards a treatment work. It is situated in a triangle formed by railways at a level above the natural terrain, which made its construction and usage rather difficult.

The basin consists of two compartments: the first, frequently used, was designed to expedite cleaning after each filling and the second, fed by overflow from the first, is rarely used. The ensemble is supervised and managed by remote control from a central post.

---

*official ϕ level for topography.

Stormwater basin   **BEQUIGNEAUX**
Institutional data

Owner: Urban Community of Bordeaux (CUB)
Project supervisor: Directorate of Technical Services, Urban Community of Bordeaux
Operator: Lyonnaise des Eaux-Dumez

Financial data

Year of construction: 1987   Duration of construction work: 18 months
Total cost (Francs, 1993, without taxes): 19.2 MF
Part contributing to hydraulic protection:–
Part specially provided for ancillary uses:–

Technical data

Associated catchment area:–   Surface area of basin: 5.5 ha
Runoff coefficient:–   Storage volume: 100,000 m^3
Drawdown range 10 years:–   Depth: –
Flow-control device: trapezoidal weir + automated valves
Regulated flow rate:–
Phreatic (free) groundwater: no   Watertightness: no
Receiving waters: Lake Bordeaux
Pretreatment: grit chamber
Dyke: yes
Special features: compartmentalisation to facilitate management; the most polluted water diverted to a treatment work; the remaining water flows into Bordeaux Lake
Major problems encountered: stability of dyke, seepage and percolation

DRY BASIN OF BRON

Commune of Bron (Rhône)

Photo, courtesy of Agence de l'Eau Rhône-Méditerranée-Corse

The dry basin of the Triangle de Bron receives stormwater from a zone which is essentially a centre of commercial activities and thus has large impermeable parking areas. This business zone is situated in a plain that is not very permeable, and is also distant from natural outlets. The basin is drained by constant discharge to a 'mini' stormwater network, which eventually carries water to an infiltration basin.

Stormwater basin   **BRON**

Institutional data

Owner: Society of Utilities of Lyonnaise Region
Project supervisor: Director of Water Department, COURLY (Urban Community of Lyon)
Operator: Director of Water Department, Courly

Financial data

Year of construction: 1982   Duration of construction work: 3 months
Total cost (Francs, 1993, without taxes): 1.2 MF
Part contributing to flood protection:–
Part specially provided for ancillary uses:–

Technical data

Associated catchment area: 30 ha   Surface area of basin: ha
Effective volume 20 years: 3800 m^3 Storage volume: 5600 m^3
Drawdown range 10 years: cm   Mean depth: m
Flow-control device: Hydroslide   Regulated flow rate: 300 l/s
Phreatic (free) groundwater: yes   Watertightness:–
Receiving waters: groundwater (infiltration) + drainage network + Rhône river
Pretreatment: screens upstream
Dyke:–
Special features: commercial area, tertiary activities and hotels
Major problems encountered:–

UNDERGROUND TANK OF REMICOURT

Commune of Villers-lès-Nancy (Meurthe-et-Moselle)

This tank is located in a restricted part of the park of Remicourt chateau. Stormwater transits through two storm-sewers 800 mm and 1600 mm in diameter and falls into the first compartment (82 × 12 m). The second compartment is filled by the overflow that crests the longitudinal separation weir. Compartment 2 empties progressively through a pipe 500 mm in diameter, equipped with a static valve.

Telesurveillance is effected by ultrasonic limnimeters.

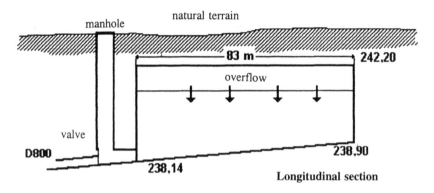

Longitudinal section

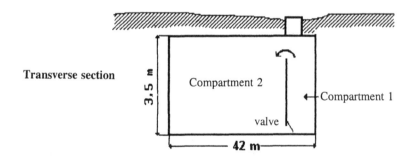

Transverse section

Stormwater basin   **REMICOURT**
Institutional data

Owner: Urban District of Nancy
Project supervisor: Urban District of Nancy
Operator: Urban District of Nancy

Financial data

Year of construction: 1989–90   Duration of construction work: 8 months
Total cost (Francs, 1993, without taxes): 7.1 MF
Part contributing to flood protection: 100%
Part specially provided for ancillary uses: nil

Technical data

Associated catchment area: 107 ha   Surface area of basin: 3500 m^2
Coefficient of impermeabilisation: 32%   Storage volume: 11,000 m^3
                                            Mean depth: 3.5 m
Flow-control device: 'Hydroslide' valve 400 mm diam
Regulated flow rate: 200 l/s
Phreatic (free) groundwater:–   Watertightness:–
Receiving waters:–
Pretreatment: grit chamber
Dyke: no
Special features: 4 vent stacks for aeration with anti-odour devices (0.80 × 0.80 m)
Major problems encountered: unaesthetic appearance of work situated in centre of the city;
need for prohibiting access

UNDERGROUND TANK OF BERGONIE

Commune of Bègles (Gironde)

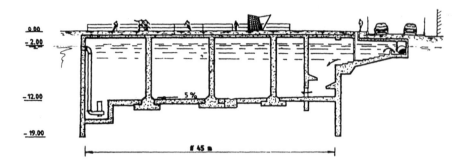

The Bergonié underground tank was planned to provide relief to the pumping station of Noutary by diverting part of the 7.1 m³/s peak discharge.

Constructed underneath a stadium, this cylindrical reservoir of 45 m diameter can store 16,000 m³, corresponding to a water depth of 10 m. Evacuation is done by pumps (2 submersible sets, each capable of pumping out 500 l/s).

The moulded wall descends 19 m, i.e., terminates 7 m below the floor. The residual peak storm discharge is estimated to be 300 l/h, corresponding to a water level of 10 m above the floor.

Stormwater basin   **BERGONIE**

Institutional data

Owner: Urban community of Bordeaux
Project supervisor: Lyonnaise des Eaux-Dumez
Operator: Lyonnaise des Eaux-Dumez

Financial data

Year of construction: 1992-93   Duration of construction work: 1 year
Total cost (Francs, 1993, without taxes): 21.4 MF
Part contributing to flood protection:–
Part specially provided for ancillary uses:–

Technical data

Associated catchment area: ha   Surface area of basin:–
Runoff coefficient:–   Storage volume: 16,000 m^3
Drawdown range 10 years: cm   Mean depth: 10 m
Flow-control device: pumps   Regulated flow rate: 1 m^3/s
Phreatic (free) groundwater: yes   Watertightness: yes
Receiving waters: collector drains
Pretreatment:–
Dyke: no
Special features: very good performance/cost ratio; cylindrical basin, precast concrete walls anchored in impermeable substratum; floor drains
Major problems encountered:–

UNDERGROUND TANK OF PERINOT

Commune of Bordeaux (Gironde)

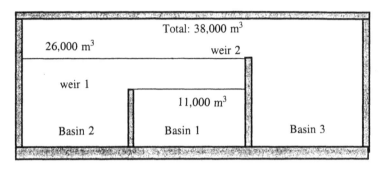

Planned to combat inundation and also to optimise management of the drainage system, this retention system was located on a diversion of the drainage network of Caudéran. It is fed only during heavy rainstorms.

Situated in a very urbanised zone, this tank is in complete harmony with the site and the environment because it lies beneath playgrounds and all appurtenances, including equipment and operational housings and the shaft access, are completely integrated in the sports complex building.

Like many other basins of the Urban Community of Bordeaux, this unit, too, is controlled round-the-clock from a central post; the level gauges in the tank and the sewerage drains continuously relay information regarding the status quo. Valves at the inlet and outlet are operated by remote control.

Storm run-off basin   **PERINOT**

Institutional data

Owner: Urban Community of Bordeaux (CUB)
Project supervisor: Lyonnaise des Eaux-Dumez
Operator: Lyonnaise des Eaux-Dumez

Financial data

Year of construction: 1983   Duration of construction work: 1 year
Total cost (Francs, 1993, without taxes): 34.5 MF
Part contributing to flood protection: 100%
Part specially provided for ancillary uses: nil

Technical data

Associated catchment area: 217 ha   Surface area of basin: 1 ha
Runoff coefficient:–   Storage volume: 38,000 m^3
          Depth: 4 m
Flow-control device: automated valves   Regulated flow rate:–
Phreatic (free) groundwater: no   Watertightness: yes
Receiving waters: Caudéran river
Pretreatment: no
Dyke: no
Special features: compartmentalisation; tank overtopped by a lawn and playgrounds; tele-surveillance and remote contol
Major problems encountered: vandalism via aeration vent stacks

UNDERGROUND TANK OF MARE AUX POUTRES

Commune of Sevran (Seine-Saint-Denis)

The tank of Mare aux Poutres is situated on the catchment area of the Morée drain and is a part of the programme for combating inundations in the north-eastern part of the district. It is built in two parts: one underground and used for storage and sedimentation, the other, an area prone to inundation, grass-covered and used as a football field. It is fed by pumps located in the first compartment. Its hydraulic management is locally automated and equipped with telesurveillance; remote control is possible during the storm period.

The underground tank is cleaned by means of hanging flush cisterns fed by city water.

Stormwater basin   **MARE AUX POUTRES**
Institutional data

Owner: Council General of Seine-Saint-Denis
Project supervisor: Director of Water and Sanitation
Operator: Council General

Financial data

Year of construction: 1992-93   Duration of construction work: 2 years
Total cost (Francs, 1993, without taxes): 25 MF
Part contributing to flood protection: 50%
Part specially provided for anciliary uses: 50%

Technical data

Associated catchment area: 133 ha   Surface area of basin: 1 ha
Filled volume 10 years: 10,000 m^3 Storage volume: 18,500 m^3
Drawdown range 10 years: cm   Mean depth: m
Flow-control devices: weir and valves   Regulated flow rate: 1 m^3/s
Phreatic (free) groundwater: yes   Watertightness: –
Receiving waters: sewer for decanted water; drain for water-sludge from the bottom
Pretreatment: flow-breaker to remove coarse matter
Dyke: no
Special features: moulded walls; 10,000 m^3 storage in zone prone to inundation; 8500 m^3 in underground compartment
Major problems encountered: groundwater seepage

References

1. BACHOC, A. and CHEBBO, G. (1992). Caractérisation des solides en suspension dans les rejets pluviaux urbains. Actes des 3es journées du DEA, Sciences et Techniques de l'Environnement, Paris.
2. BACHOC, A., CHOCAT, B. and COGEZ, C. (1992). Hydrologie urbaine, les bassins nouvelle vague. Colloque de Pantin.
3. BERGUE, J.M. (1979). Les bassins de retenue d'eaux pluviales. Synthèse bibliographique. STU, Ministère de l'Environnement et du Cadre de Vie, 70 p.
4. BERGUE, J.M., DEUTSCH, J. C. and CHERON, J. (1980). Guide technique des bassins de retenue dÆeaux pluviales. STU (unpublished).
5. BILLECOCQ, M. and SIMON, H. (1979). Les retenues d'eaux pluviales, conclusions d'une étude comparative sur 5 bassins. TSM Eau no. 1, pp. 41–46.
6. BOUCHET, M., DEUTSCH, J. C. and BILLECOCQ, M. Etude des possibilités d'auto-épuration des bassins de retenue. Cas de l'étang de Ville Evrard. DDE-IRCHA, 94 p.
7. Deleted.
8. BOYER, C. (1993). Méthodologie d'aide à la décision pour les projets d'assainissement. Rapport CERGRENE.
9. CAVELIER JACQUEMAIN, C. (1977). Traitement du bassin de Sainte Suzanne par la craie en poudre. Contribution à l'étude d'un moyen de lutte contre l'eutrophisation. Thèse Doct. Univ. Rennes, 244 p.
10. CHEBBO, G. (1992). Solides des rejects pluviaux urbains; caractérisation et traitabilité. Thèse ENPC.
11. CHOCAT, B., SEGUIN, D. and THIBAULT, S. (1986). Hydrologie urbaine et assainissement urbain. INSA (cours).
12. COGEZ, C. (1990). Le diagnostic hydrologique préalable à toute opération d'aménagement. ENPC, session de formation, 15 p.
13. Comité Français des Géotextiles (1982). Recommandations générales pour la réception et la mise en oeuvre des géotextiles.
14. Comité Français des Géotextiles et des Géomembranes (CFGG). (1990). Projet d'étanchéité des ouvrages hydrauliques par géomembranes.
15. Conseil Régional d'Ile-de-France, Agence de l'Eau Seine-Normandie, Direction Régionale de l'Equipement (1990). Esquisse d'un programme d'action pour l'assainissement des eaux pluviales en Ile-de-France, 19 p.
16. Conseil Général de Seine-Saint-Denis (1992). Les bassins nouvelle vague. Hydrotechnologie urbaine. Colloque sur les bassins de retenue, 16–18 juin 1992, 324 p.
17. CORONO, G. and MURET J.P. (1978). Loisirs - Guide pratique des équipements. CRU, 713 p.
18. COSTE, C. and LOUDET, M. (1987). L'assainissement en milieu urbain ou rural. Editions du Moniteur, volume 1, 233 p: volume 2, 265 p.
19. CRESTIN, J.P., ALQUIER, M. and DARTUS, D. (1982). Amélioration des performances des décanteurs, modèle mathématique. CNRS.
20. CUB-Lyonnaise des Eaux (1985). Gestion automatisée d'une retenue d'étalement.

21. DEE Seine-Saint-Denis. Les retenues d'eaux pluviales, étude comparative sur 5 bassins, 105 p.

22. DE SABLET, M. (1988). Des espaces urbains agréables à vivre. Ed. du Moniteur, 255 p.

23. DEGARDIN, P. (1991). Quantification des rejets urbain de temps de pluie en région parisienne. AESN.

24. DELFAUT, A., JARDIN, J. and BALDIT, R. (1984). Constatations sur la digue de Maurepas-Courance. Bull. liaison Labo. P & C no. 131, 7-22 pp.

25. DEMOUCHY, G. (1990). Bassin d'orage de la ville nouvelle de Vitrolles. Bull. SHF no. 7, 5 p.

26. DESBORDES, M., DEUTSCH, J.C. and FREROT, A. (1990). Les eaux de pluie dans la ville. La Recherche no. 221.

27. DRM-STU (1991). Réconcilier l'eau et la ville par la maîtrise des eaux pluviales.

28. DUBSOIS, M. and JACOBSEN, E. (1988). Bassins de retenue: pièces d'eau ou cloaques? TSM Eau no. 10, 513–516.

29. DUPUY, G. and KNAEBEL, G. (1982). Assainir la ville hier et aujourd'hui. Dunod, 91 p.

30. DUSSART, B. (1986). Limnologie, Etude des eaux continentales. Gauthier-Villars, 677 p.

31. DUTARTRE, A., GRILLAS, P. and LEVET, D. (1990). Revue sur les techniques de contrôle des plantes aquatiques. 14e conf. du COLUMA, Versailles pp. 265–273.

32. DUTARTRE, A. and TREMEA, L. (1990). Contrôle mécanique des plantes aquatiques. 14e conf. du COLUMA, Versailles, pp. 275-282.

33. EHRHARDT, J.P. and SEGUIN, G. (1978). Le plancton: composition, écologie, pollution. Gauthier-Villars.

34. ENPC (1996). Evacuation des eaux pluviales.

35. Etude inter-Agences, LCPC (1988). Prise en compte des conditions géotechniques dans les projets de canalisations d'assainissement.

36. FREROT, A. (1980). Etude de faisabilité de la modélisation dynamique pour la gestion automatisée des réseaux d'assainissement. STU.

37. FREROT, A. (1987). Procédures d'optimisation des consignes de gestion d'un réseau d'assainissement automatisé. Thèse ENPC, 296 p. + annexes.

38. GUITTARD, M. (1978). Exécution et entretien des bassins de retenue. ENPC (cours).

39. HIRSCHAUER, A. (1991). Les bassins de retenue d'eaux pluviales. Incidences de l'environnement géotechnique sur leur conception. Conseil Général de Seine Saint-Denis, 244 p.

40. Instruction technique relative aux réseaux d'seaux d'assainissement des agglomérations (1977). Circulaire no 77.284/INT. Imprimerie Nationale.

41. JOSSEAUME, H. (1983). Recommandations pour l'étude des digues et barrages en terre de faible hauteur. Doc. interne LPC.

42. JOSSEAUME, H. and KADHAVI, C. (1984). Interprétation des measures de pression interstitielle dans la digue de Maurepas-Courance; anisotropie de perméabilité de l'ouvrage. Bull. liaison Labo P & C, pp. 7–22.

43. LCPC-SETRA (1992). Réalisation des remblais et des couches de forme; guide technique.

44. LCPC-SETRA (1987). Méthode de terrassements routiers utilisée en France.

45. LCPC-SETRA (1988). Le déroctage à l'explosif dans les travaux routiers; guide technique.

46. LEROY, R. (1980). La prolifération des végétaux aquatiques. Mesures curatives. Bull. Liais. Labo P + C no. 107, pp. 27–36.

47. LHM, Labo, Botanique Université Bordeaux (1982). Les bassins de retenue d'eaux pluviales, glossaire descriptif des indicateurs biologiques.

48. LUCCHETTA, J.C. (1985). Etude de l'influence du gypse micronisé sur l'eau, fonds d'étangs et rivières. Bull. Liais. CSP Compiègne, pp. 11–18.

49. MAES, C. (1992). Les contraintes techniques et administratives relatives aux contraintes d'insertion dans un site urbain des bassins de dépollution des eaux pluviales. ENGREF, 63 p. + annexes.

50. MALET, M. and TA, T.T. (1985). Techniques alternatives en assainissement pluvial; impact sur le milieu social et sur l'environnement. Rapport de synthèse. Plan Urbain. 80 p.

51. MARCHAND, A., BADOT, R., de BELLY, B. and ROMAIN, M. (1993). Les bassins de rétention des eaux pluviales. Mode d'emploi. 20 ans d'expérience au District Urbain de Nancy. Nancie, 222 p.

52. MARTE, C. and RUPERD, Y. (1989). L'efficacité des ouvrages de traitement des eaux de ruissellement. TSM Eau no. 5, pp. 297–301.

53. Mémento technique de l'eau (1989). 2 volumes, Degremont.

54. Ministère de l'Agriculture (1977). Technique des barrages en aménagement rural. CEMAGREF, 325 p.

55. Ministère de l'Environnement (1992). Evaluation du risque lié au ruissellement pluvial urbain, éléments de méthode. EURYDICE 92, IPGR. 85 p.

56. Ministère de l'Equipement (1977). Qualité des eaux superficielles. Epuration. Hydrologie urbaine. Jour. d'Info. Nat. Paris.

57. Ministère de l'Equipement (1970). Moyens de lutte contre l'eutro-phisation des lacs. Etudes bibliographiques.

58. MONTEGUT, J. (1987). Les plantes aquatiques: milieu aquatique entretien, désherbage ACTA, 4 fasc.

59. NALLET, J. (1977). Traitement des bassins de retenue d'eaux pluviales à St-Quentin-en-Yvelines. Ann. TP 113, no. 1018.

60. NUMEZ VARGAS, M. (1990). Evaluation de bassins de retenue de petites tailles et autres techniques alternatives. DEA 93.

61. OLIVARY, D., MACLAIR, N. and DELBOS, B. (1990). Valorisation des plans d'eau artificiels marnants. Bas-Rhône-Languedoc, 145 p.

62. PHILIPPE, J.P. and RANCHET, J. (1987). Pollution des eaux de ruissellement pluvial en zone urbaine, synthèse des mesures sur 10 bassins versants en région parisienne. LCPC no. 142, 76 p.

63. RANCHET, J. (1985). Ouvrages de prétaitement des eaux de ruissellement routier. Annales des TP de Belgique. no. 2, pp. 142–148.

64. RANCHET, J. and RUPERD, Y. (1982). Moyens d'action pour limiter la pollution due aux eaux de ruissellement en système séparatif et unitaire; synthèse bibliographique. Rapport de recherche LCPC no. 111, 104 p.

65. REMILLARD, L. (1988). Bassin de rétention en zone urbaine. Sci. Tech. Eau 21(3): 271–276.

66. RIZET, M. (1983). Lutte contre le développement des algues par des moyens biologiques. Water Supply 1(1): 229-235.

67. ROSENBAUM, V. (1992). Plans d'eau de faible profondeur en zone urbaine. AESN-Lyonnaise-Dumez, 130, p. + bib. + annexes.

68. ROSENBAUM, V. (1992). Gestion et entretien des plans d'eau. Lyonnaise des Eaux-Dumez, 87 p.

69. ROY, B. (1985). Méthodologie d'aide à la décision multicritère. Economica, Paris.

70. RUPERD, Y. (1991). Suivi d'un séparateur lamellaire utilisé pour le traitement des eaux de ruissellement. LROP, 46 p.

71. RUPERD, Y. (1986). Efficacité des ouvrages de traitement des eaux de ruissellement. STU, 121 p.

72. RUPERD, Y. (1984). Suivi du déshuileur à plots de Bois-Robert. LROP, 45 p.

73. RUPERD, Y. (1984). Etude du fonctionnement des ouvrages de traitement des eaux de ruissellement de la Cour Rolland. DDE 78-LROP.

74. SAUVETERRE (1992). Des plans d'eau en ville. Plan Urbain, 96 p. + graphiques et photos.

75. SAUVETERRE (1985). Approche mathématique des suivis écologiques des bassins de retenue en Ile-de-France. Plan Urbain.

76. SAUVETERRE (1983). Etude économique des bassins de retenue. Agence de Bassin Seine-Normandie, 2 fasc.
77. SAUVETERRE (1981). Comportement écologique de quelques retenues d'eau pluviale. STU, 85 p.
78. SAUVETERRE (1975). Approche écologique des retenues d'eau pluviale. GCVN.
79. SETRA (1982). Recommandations pour l'assainissement routier.
80. Société du Canal de Provence (1988). Gestion centralisée du réseau de collecteurs du bassin versant de La Cadière; étude de factibilité. EPAREB, 27 p.
81. Société Hydrotechnique de France (1990). Collectif. Economie de l'hydrologie urbaine, la ville sous l'eau, l'eau sous la ville, la ville, l'eau et les sous...SHF Session no. 140.
82. Société Hydrotechnique de France (1986). Collectif. L'impact des activités humaines sur les eaux continentales. SHF, 19èmes Journ. Hydrau.
83. Société Hydrotechnique de France (1993). Collectif. La pluie, source de vie, choc de pollution. CR colloque 17–18 mars 1993. La Houille Blanche, no. spécial septembre 1993.
84. STU (1983). Migration des polluants en milieu poreux non saturé. Laboratoire d'Autun.
85. STU (1986). Guide des logiciels français en hydrologide urbaine.
86. STU (1989). Mémento sur l'évacuation des eaux pluviales. La Documentation Française, Paris, 349 p.
87. STU (1992). L'aménagement des espaces verts. Ed. du Moniteur, 292 p.
88. THOMACHOT, M. (1980). Les bassins de retenue. Doc. int. LROP.
89. VALIRON, F. and TABUCHI J.P. (1992). Maitrise de la pollution urbaine par temps de pluie. Tec & Doc-Lavoisier, 564 p.
90. VERNIERS, G., LOZE, H. and DEGEIMBRE, R. (1985). Les bassins d'orage autoroutiers: fonction, aménagement et entretein. Ann. TP Belg, no. 2, pp. 125–164.
91. Universités de montpellier et Bordeaux (1982). Glossaire descriptif des indicateurs biologiques. Méthodologies d'étude. Les bassins de retenue d'eaux pluviales, 112 p + annexes.
92. DESBORDES, M. (1977). Quelques méthodes de calcul des bassins de retenue des eaux pluviales. CEBEDEAU, 7 p.
93. STU (1982). Amélioration des performances des décanterus. Institut de Mécanique des Fluides, Toulouse, 301 p.

Glossary

albedo: fraction of the incident radiation which is diffused or reflected by a body. The albedo of a black body is zero while that of snow is 0.9

algal bloom: rising and accumulation of algae on the surface as a result of an excessive proliferation of some species in the mass of water

alternative techniques: sewerage techniques in which works other than only sewers are made use of; also called compensatory techniques or best management practices BMP.

anisotropic (environment): with heterogeneous composition along the three axes

aquifer: permeable geological formation containing a water table. This term is used particularly for environment with slit-like porosity (sand)

arrester: the embedded part of a structure, forming projection, meant to prevent sliding or oscillation of the structure on its base

Atterberg (limits of): characteristic values of a material in regard to plasticity, liquidity and shrinkage

autotrophic: capable of self-nourishment and generating organic matter from nutrients; vegetables are most frequently autotrophic

bare: materials located above a deposit of granules or rocks, laid bare before or during work on gravel or quarry

barren: bare portion under the vegetal growth

basket: geometric pile of stones maintained in a metallic or polyethylene meshwork for consolidating the banks or walls

benthic (fauna, flora): living at the bottom of a water body or watercourse

bentonite: clay of the smectite group used for making a basin watertight

berme: horizontal seat provided in an embankment for creating a break in slope, combating erosion and serving as a path etc.

biocenosis: living beings inhabitting the biotope

biomass: mass of living beings considered in lots or by systematic groups. Organic matter produced by biocenosis, useful for nourishment, fertilisation, production of energy etc.

biotope: the physical environment in which the biocenosis blossom

bloom: excessive development of algal population with eventual formation of algal bloom

catchment area: topographical surface inscribed in a closed boundary called the boundary of crest

climax: vegetal ecosystem in equilibrium with climate and other ecological factors of the biotope, not growing further in the absence of any perturbation

colline (retention tank): basin generally closed by a dyke a crossing a thalweg, meant to serve as a water reserve for agricultural and other uses

combined (network): sewerage network in which the sewers circulate both the waste water and the stormwater

dystrophic: a state of trophic unequilibrium. Hyper-eutrophication due to the excess of vegetal organic material

ecosystem: basic ecological unit formed by the biotope (the environment) and the biocenosis (inhabitants living in the biotope)

edaphic (factors): ecological factors related to soil (and not the climate)

eutrophic (environment): (etmologically) rich in nutrients

eutrophication: progressive enrichment of an environment, without disturbing equilibrium, in nutrients used primarily by autotrophic consumers but also by those depending on them. Eutrophication is a natural phenomenon but its rate of growth can increase in the case of large supply of nutrients. Eutrophication is then considered as a pollution. One should rather use the expression hyper-eutrophication

evapo-transpiration: sum of quantities of water evaporated from the soil and the water table and released by the plants

fingerlings: stocking a water body or a watercourse with young fish

geomembrane: impermeable membrane made of elastic or plastic synthetic material

geotextile: permeable synthetic textile with loose two or three-dimensional structure

helophytes: aquatic or semi-aquatic plants whose roots are submerged but other parts are exposed to the atmosphere

hydrograph: flowrate-time curve.

hydrophytes: aquatic plants embedded in silt or free, growing entirely in or on water

isotropic (environment): having homogeneous composition along the three axes

karstic (model): characteristic of calcarious regions: cavities (sink-holes, swallow-holes, gulfs) and underground water circulation (resurgences, siphons and underground rivers)

macrophytes: all plants excluding phytoplankton

marling or drawdown range: difference between the nominal level of a basin and the level attained during flooding of a given frequency

nutrients: mineral elements essential for the growth of autotrophic (vegetal) organisms

oedometry: relating to the study and measurement of compression of embankments

oligotrophic (environment): poor in nutrients

oro-hydrophic (maps): in which only relief and watercourses are shown

photosynthesis: synthesis by green plants of fundamental biochemical compounds from water and mineral salts, by making use of the ability of the chloroplasts to capture a part of the solar energy

phytophagic: eating plants

piezometer: a device for measuring or recording the water table level

plankton: ensemble of micro-organisms living in water, constituted by phytoplankton and zooplankton

Proctor: compactness test of a material with a prescribed energy, enabling the tracing of the curve of variation of density with the quantum of water of constitution

puddle: a clay or compacted concrete layer used for coating the bottom and walls of a basin

recess: space provided between the crest side of a dyke or a bank of canal and the highest water point (PHE)

revised (plans): plans generally chalked out after taking into account the actually completed portion of work

ring/hoop: metallic ring or belt used for reinforcing wooden pieces

rot-resistant: tropical wood, practically rot-proof, widely used for protecting banks against erosion

scumboard: wall forming a barrage for grease and floating matter

step: projection made on a surface for retention on it of a material or a structure

stiffener: a sleeper made of concrete, metal or wood, placed between two walls to prevent their drawing closer under the force of earth (or water) applied on their outer faces

strand: a trench made in a rock or brick-work for anchoring a structure

separate (network): network consisting of separate sewers for waste water and stormwater

quicksand: slit or hole in a dyke or an embankment through which water from a basin or a canal flows

thalweg: boundary of the bottom of a valley, eventually followed by a watercourse

thermocline: zone of discontinuity between two water masses at different temperatures

tropic: concerning the nutrition of biocenosis

trophic (chain): ensemble of organisms which ensure successively the transfer of material. A trophic chain always includes the producers (plants), consumers (herbivorous, carnivorous, omnivorous) and decomposers (detritivorous and coprophagous)

trophic (level): an element of the trophic chain

beams/rods: connecting beams, generally of wood, fixed vertically, maintaing a bank by preventing its sliding and erosion

ubiquitous: omnipresent

water stream: lower generator of a pipe or a conduit

weeding: seasonal cutting and thinning of aquatic plants

Index

Milton Keynes UK
Ingram Content Group UK Ltd.
UKHW031135141024
449569UK00006B/177